HOUGHTON MIFFLIN HARCOURT

Math

Intensive Intervention

Skill Packs

Grade 2

www.hmhschool.com

Printed in the United States of America

ISBN 13: 978-0-15-377027-2
ISBN 10: 0-15-377027-9

4 5 6 7 8 9 10 0982 17 16 15 14 13 12 11 10

Contents

Skills

Name________________________________

Use Pictures to Add

Skill 1

Learn the Math

You can join groups to find how many there are in all.

4 apples and 2 apples

Count all the apples. 4 and 2 is __6__ .

_____ peaches and _____ peach

Count all the peaches. 3 and 1 is _____ .

Write how many there are in all.

1.

2 and 3 is 5.

2.

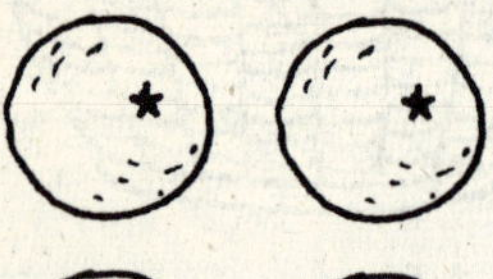

4 and 1 is ______.

3.

1 and 1 is ______.

4.

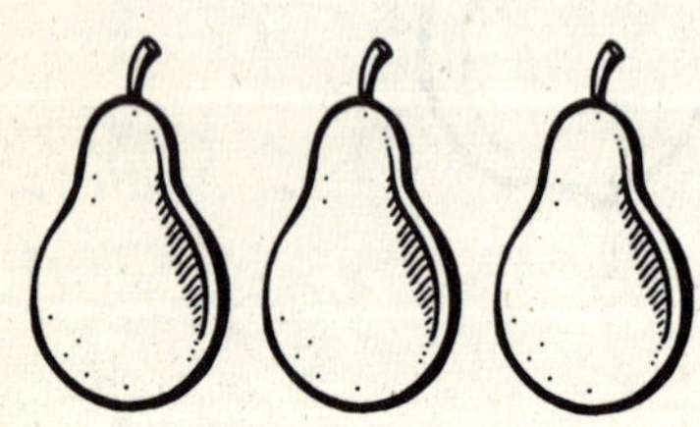

3 and 1 is ______.

5.

2 and 1 is ______.

Name ______________________________

Use Symbols to Add

Skill 2

Learn the Math

You can write an addition sentence with + and = .

Vocabulary

plus (+)
is equal to (=)
sum
addition sentence

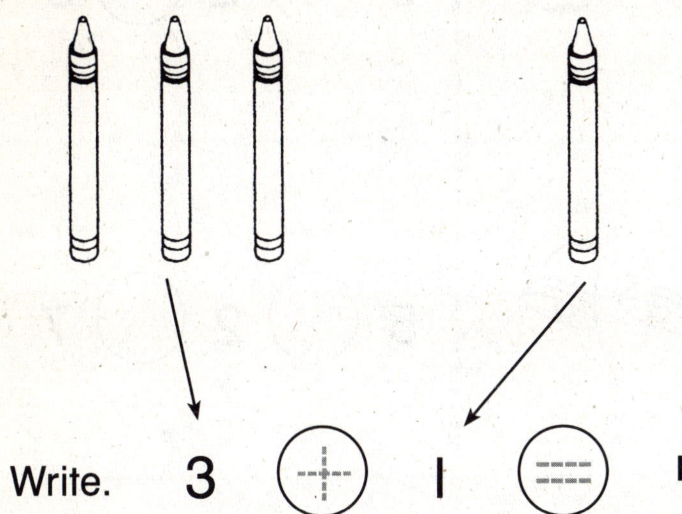

Write. 3 (+) 1 (=) 4

Say. 3 plus 1 is equal to 4.

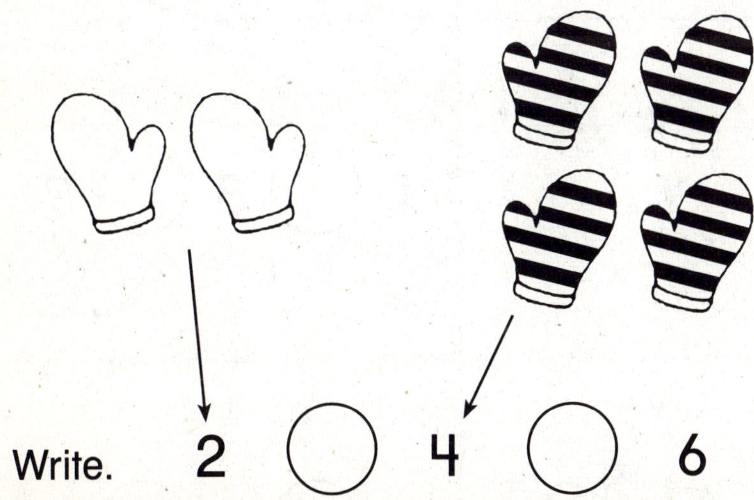

Write. 2 ( ) 4 ( ) 6

Say. 2 plus 4 is equal to 6.

Use the picture. Write the addition sentence.

1.

3 + 2 = 5

2.

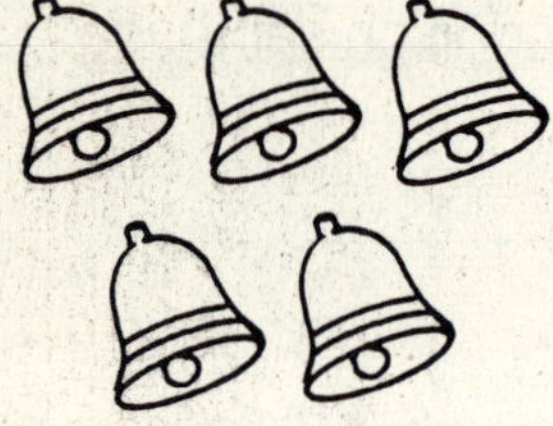

5 ◯ 2 ◯ 7

3.

2 ◯ 2 ◯ 4

4.

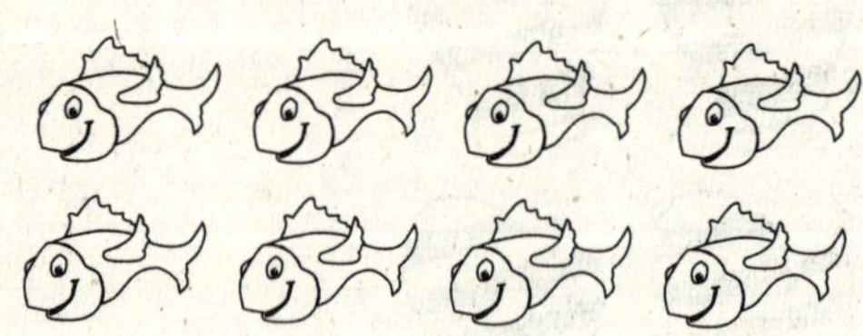

1 ◯ 8 ◯ 9

5.

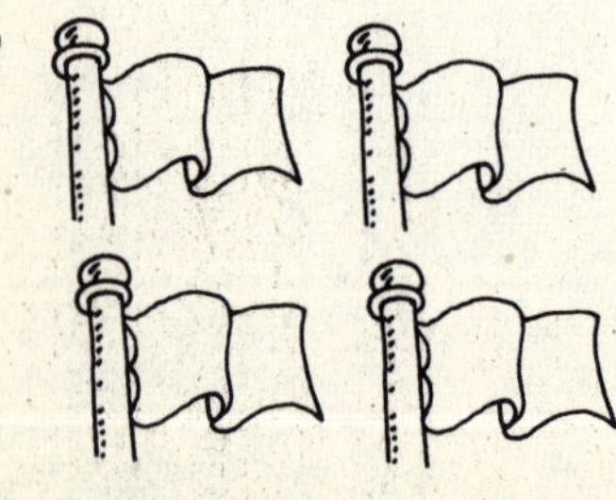

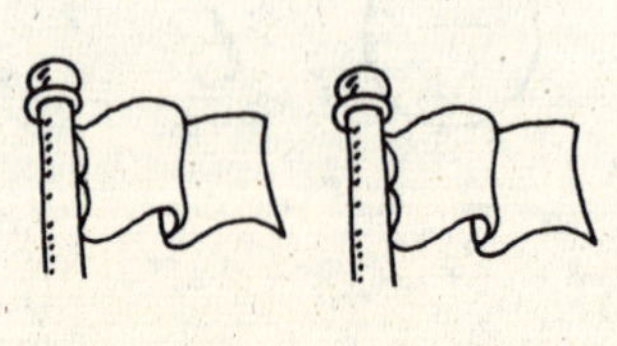

4 ◯ 2 ◯ 6

Name ____________________

Learn the Math

You can find how many are left.

4 rabbits take away 1 rabbit

Count the rabbits that are left.

4 take away 1 is 3.

_____ stars take away _____ stars

Count the stars that are left.

5 take away 2 is _____.

Write how many are left.

1.

6 take away 2 is 4.

2.

5 take away 1 is ______.

3.

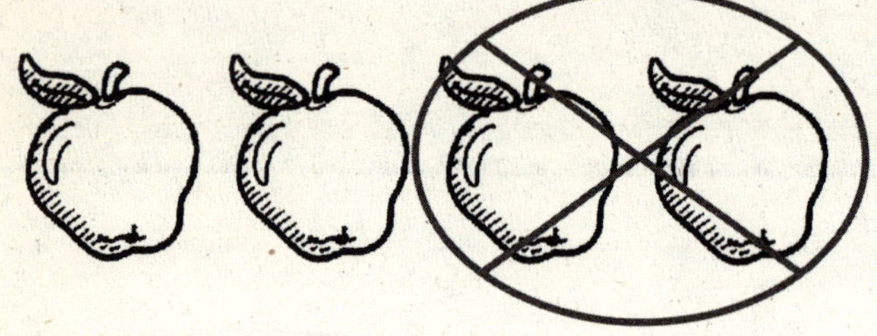

4 take away 2 is ______.

4.

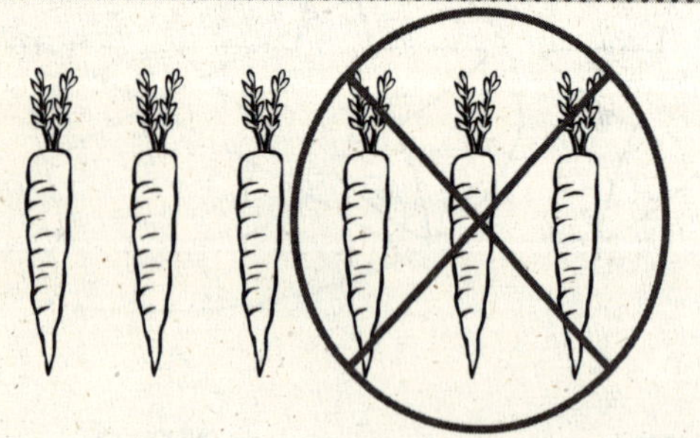

6 take away 3 is ______.

5.

5 take away 3 is ______.

Name ____________________

Use Symbols to Subtract

Skill 4

Learn the Math

You can write a subtraction sentence with − and =.

Vocabulary
minus (−)
is equal to (=)
subtraction sentence

Write. 4  2  2

Say. 4 minus 2 is equal to 2.

Write. 5  1  4

Say. 5 minus 1 is equal to 4.

Do the Math

Complete the subtraction sentence.

1\. 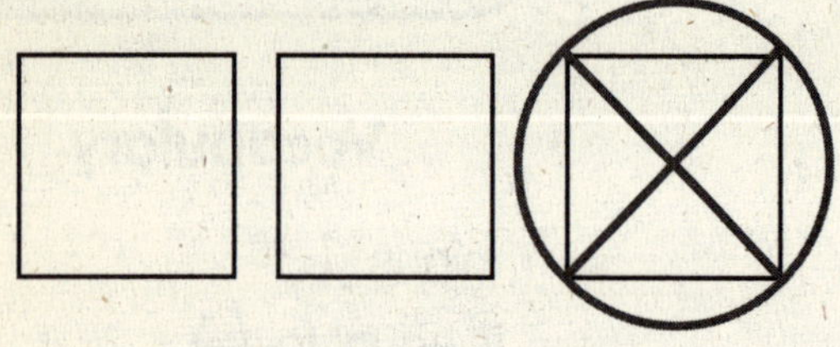

3  1 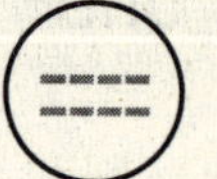 2

2\. 

6 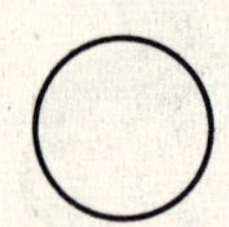 3 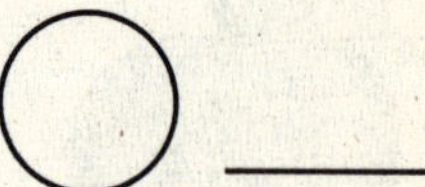

3\. 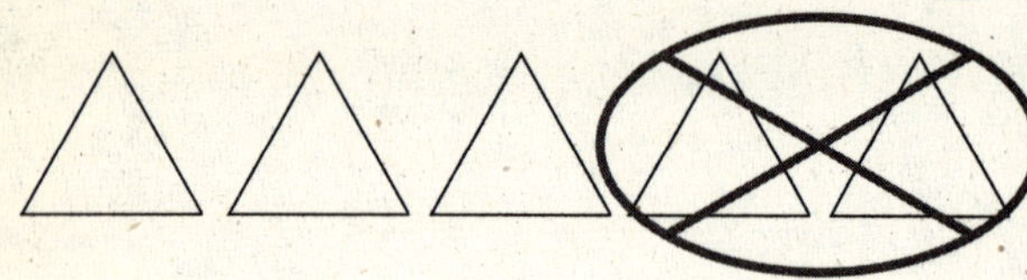

5 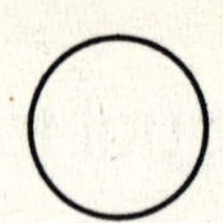 2 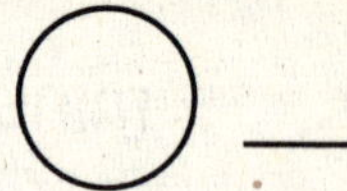

4\. 

7 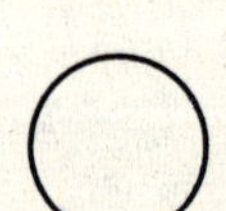 4 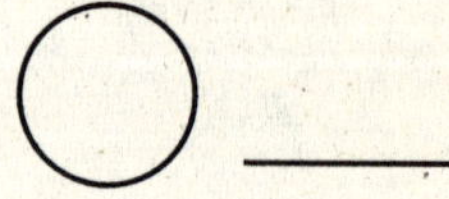

5\. 

10  5 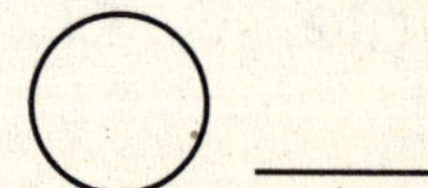

Name______________________________________

Learn the Math

You can use counters to show numbers.

How can you show 6 with counters?
Draw counters to show 6.

Draw counters to show 9.

Draw counters to show 14.

Draw counters to show the number.

1. 8

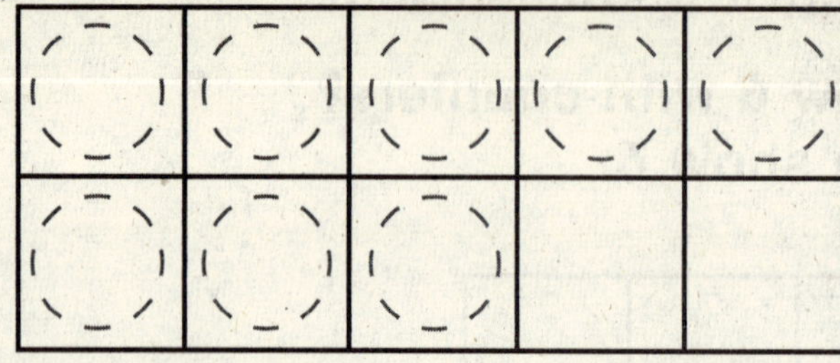

2. 12

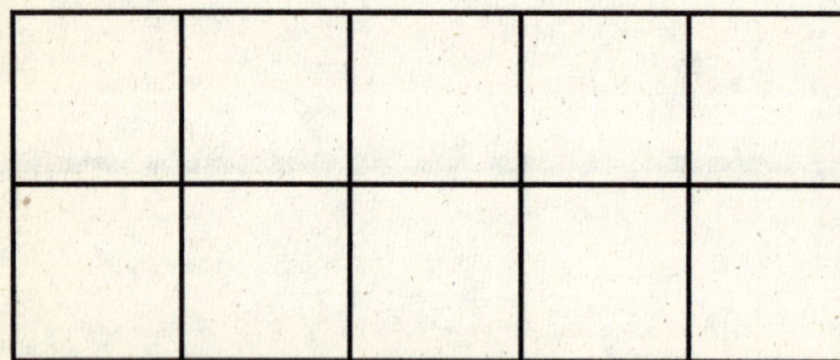

3. 15

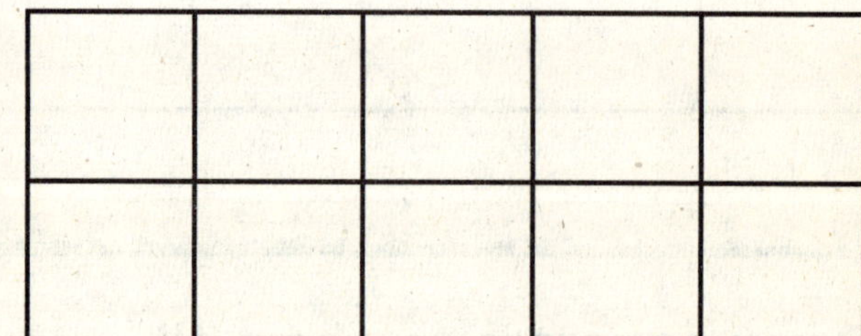

4. 7

Name ______________________________

Identify Numbers to 30

Skill 6

Learn the Math

You can use groups of 10 to find how many.

How many ☾ are there?

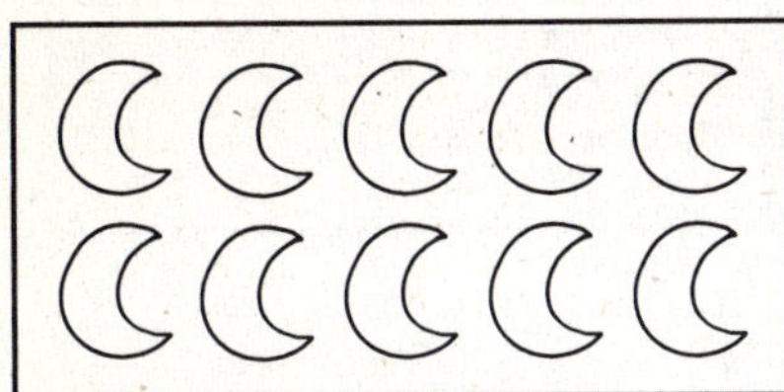

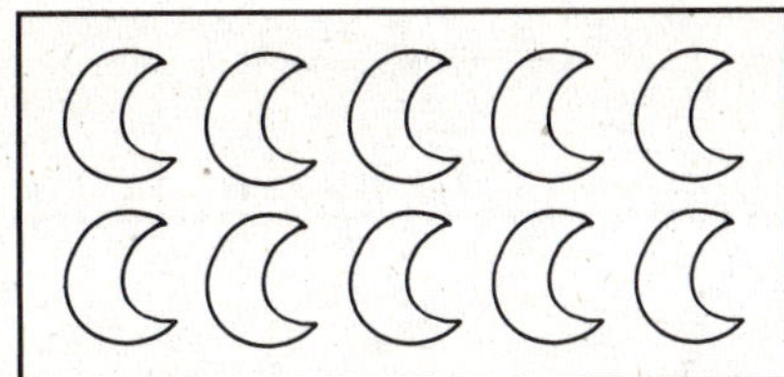

Count groups of 10 first.

How many groups of 10 ☾ are there? 2   2 tens = 20

How many other ☾ are there? 4   4 ones = 4

How many ☾ are there in all? 24   2 tens 4 ones = 24

How many ERASER are there?

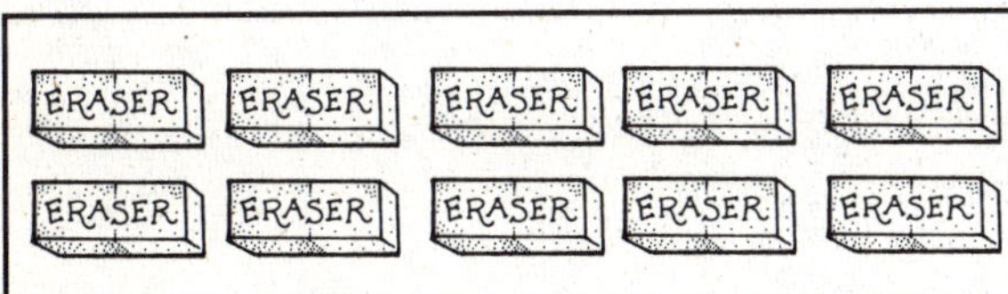

Count the group of 10 first.

Then count the rest.   1 ten = ____

There are ______ ERASER in all.   5 ones = ____

1 ten 5 ones = ____

Write how many.

1.

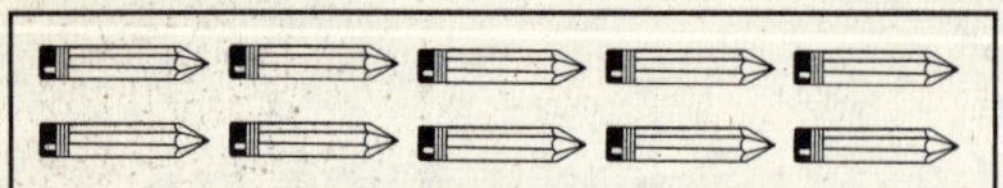

15

2.

____

3.

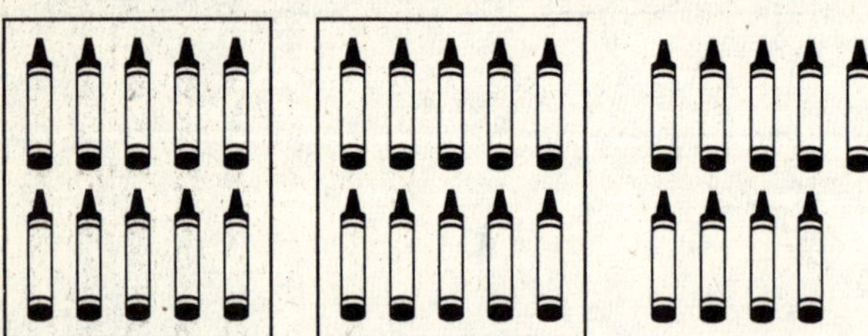

____

4.

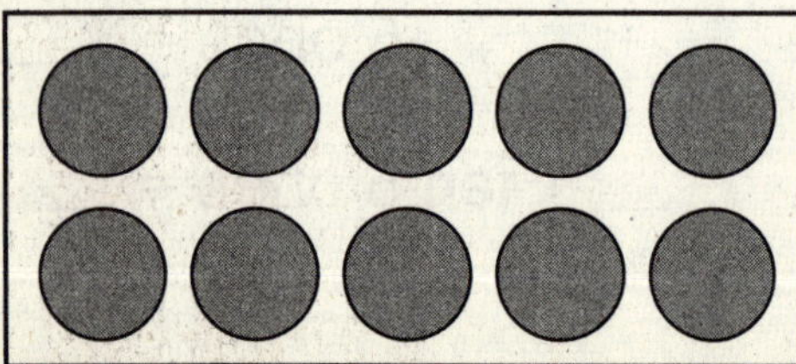

____

5.

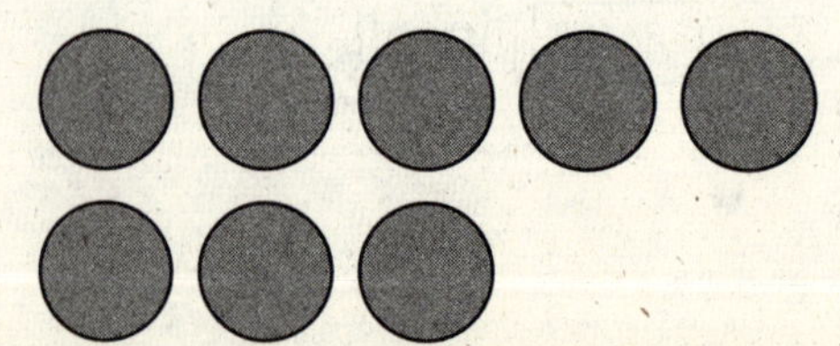

____

Name____

Compare Numbers to 30

Skill 7

Vocabulary

tens

ones

Learn the Math

Find the greater number.

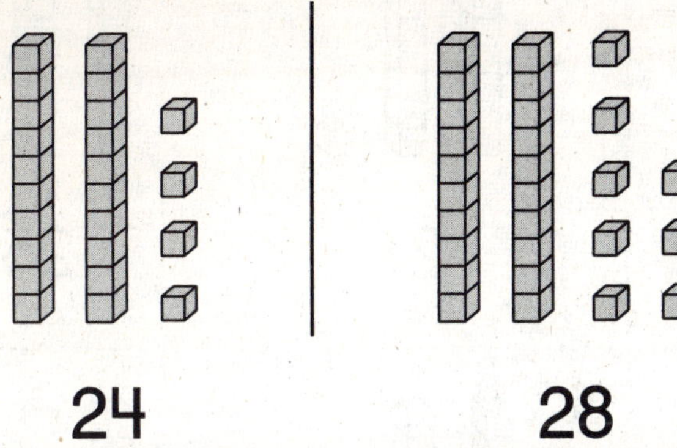

24 28

Step 1 Compare the tens.

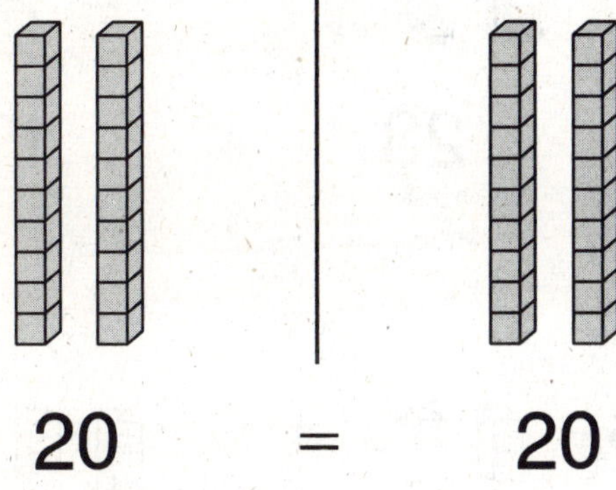

20 = 20

Step 2 If the tens are the same, compare the ones.

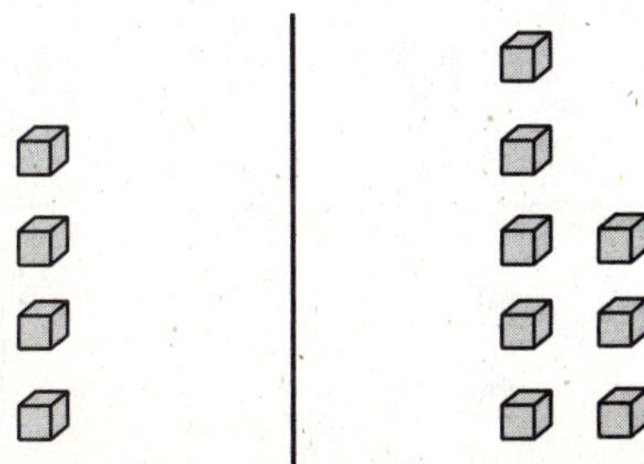

4 is less than 8

So, 28 is the greater number.

Compare. Circle the greater number.

1. 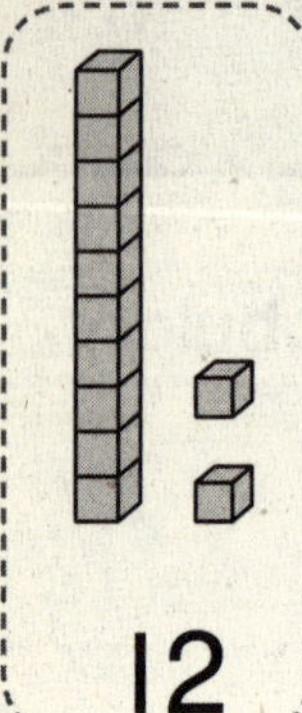 12 11

2. 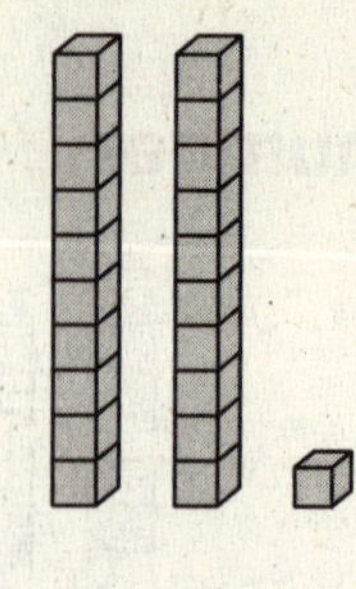 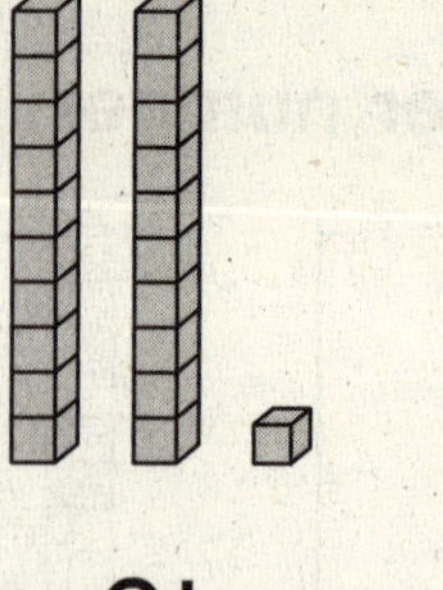 21 25

3.  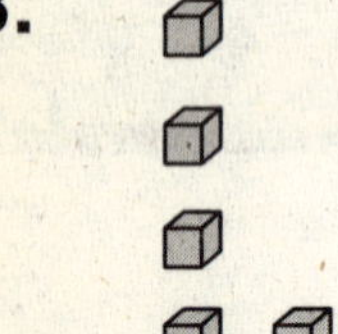 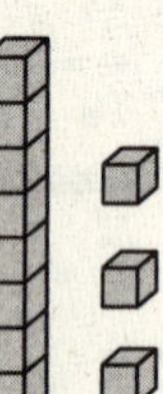 7 14

4. 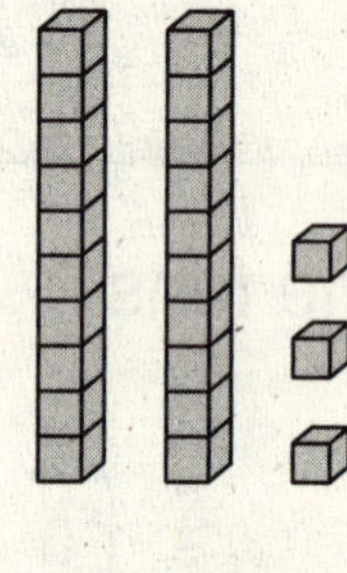 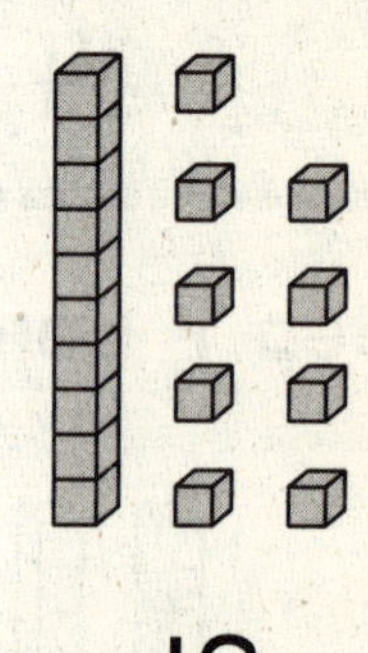 23 19

Compare. Circle the lesser number.

5. 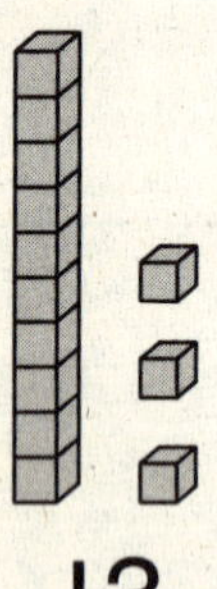  13 5

6. 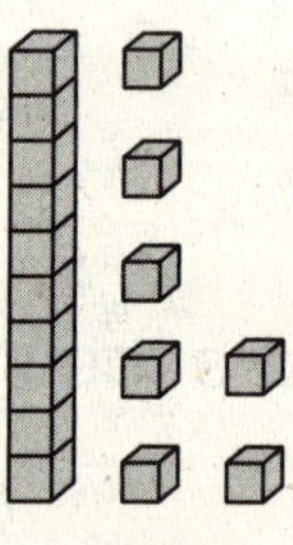 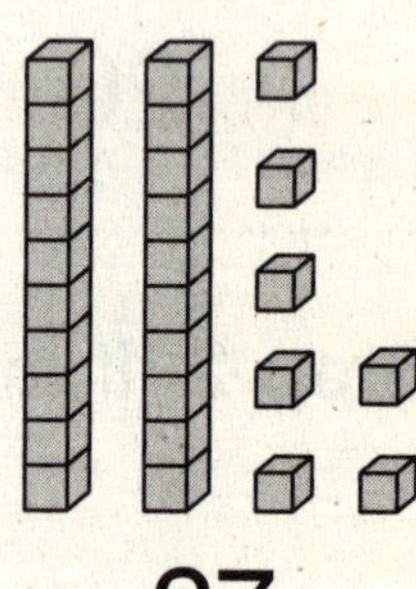 17 27

7. 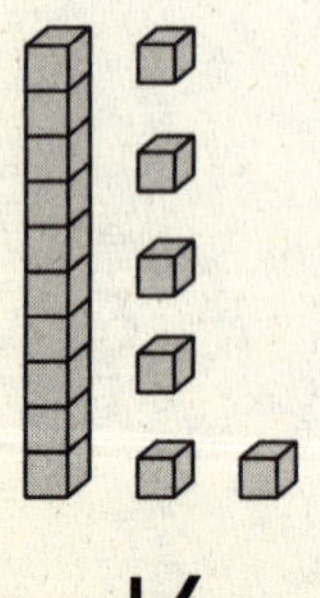  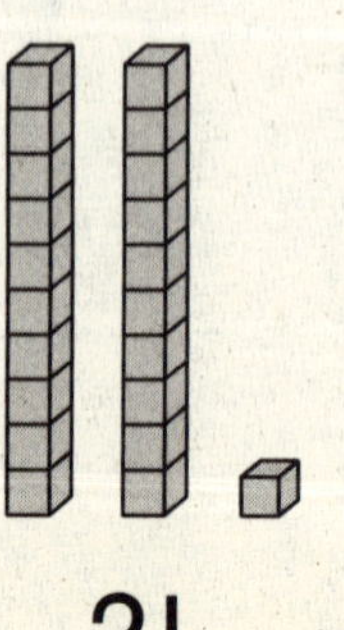 16 21

8. 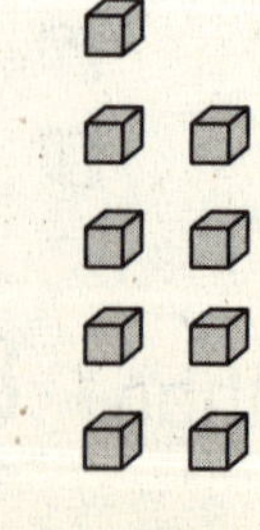 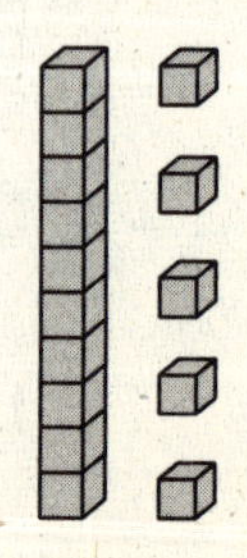 9 15

Name ____________________

Order on a Number Line

Skill 8

Learn the Math

You can use a number line to find the number that is before, after, or between other numbers.

Vocabulary

- before
- between
- after
- number line

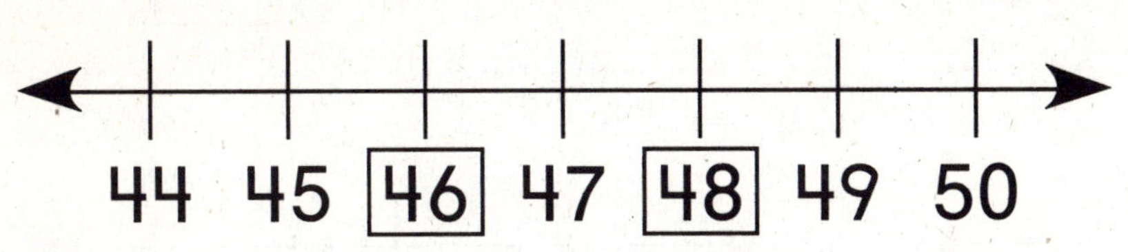

What number is just before 46?
Find 46.

45 is just before 46.

What number is just after 48?
Find 48.

49 is just after 48.

What number is between 46 and 48?
Find 46 and 48.

____ is between 46 and 48.

What number is just after 24?

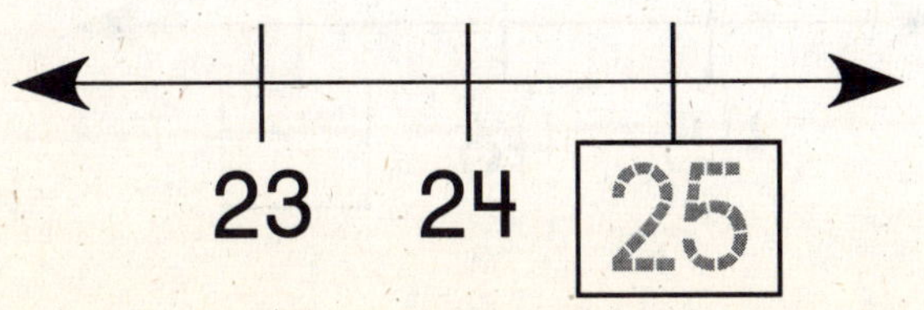

____ is just after 24.

Write the number that is just before, just after, or between.

1.

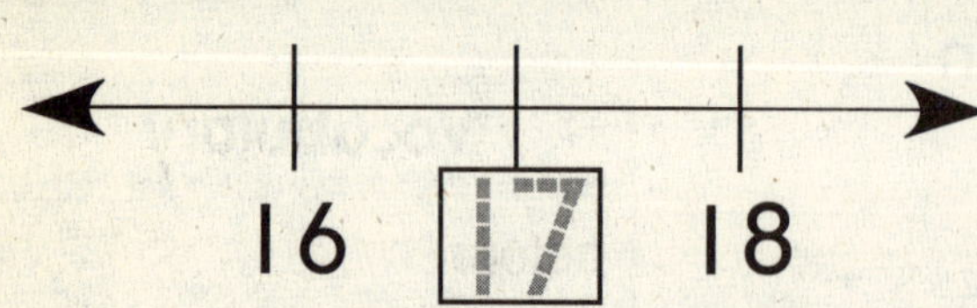

2.

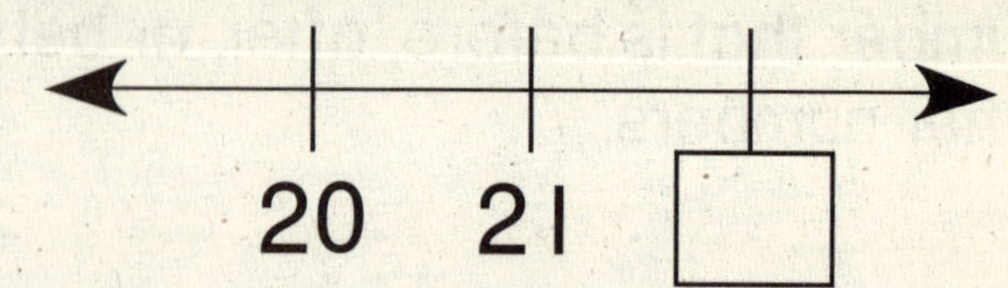

3.

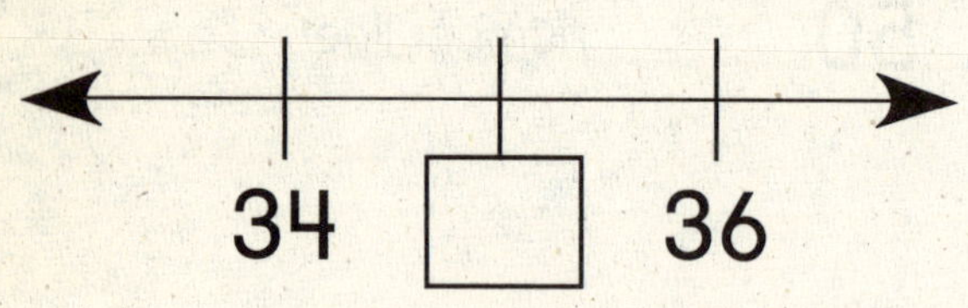

4.

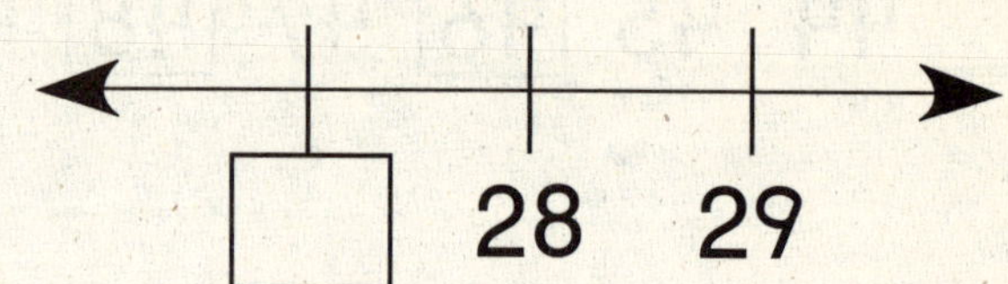

5.

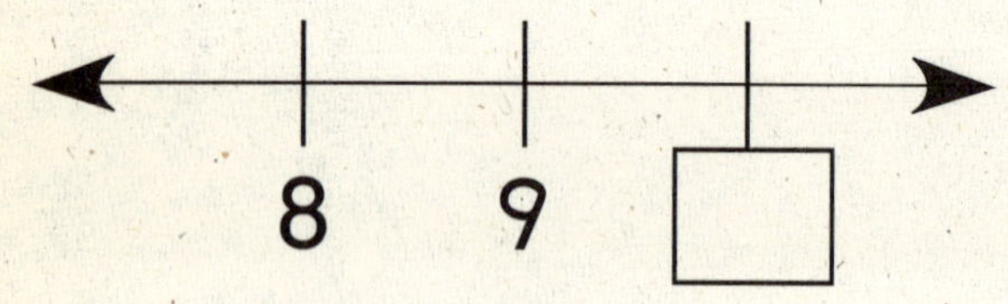

6.

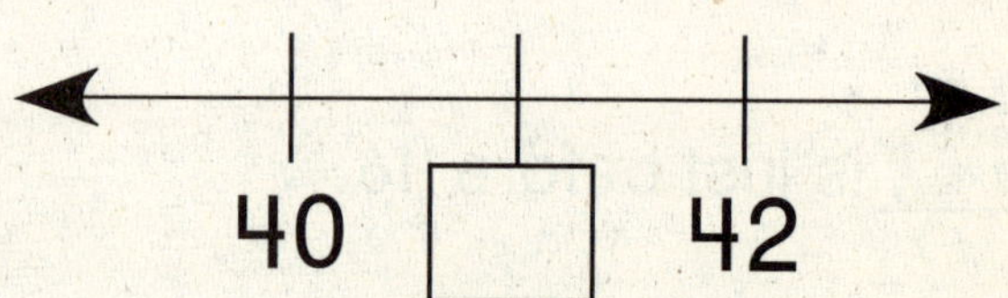

7.

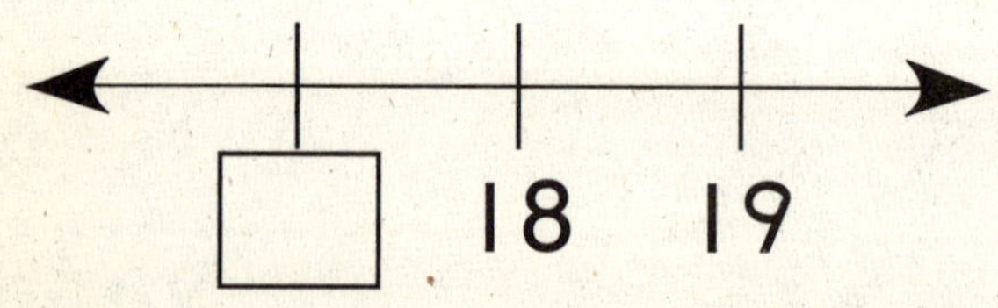

8.

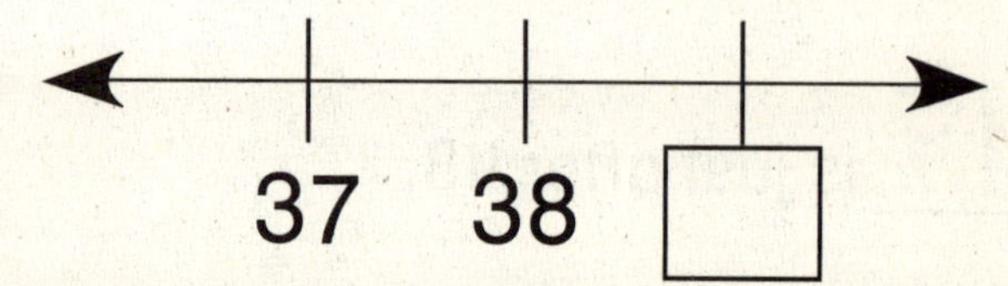

9.

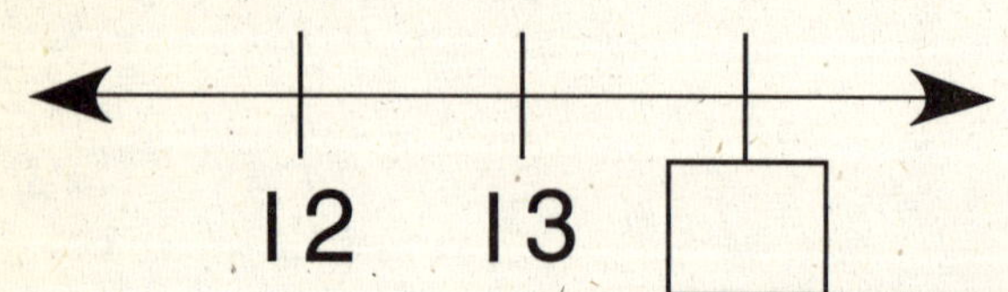

10.

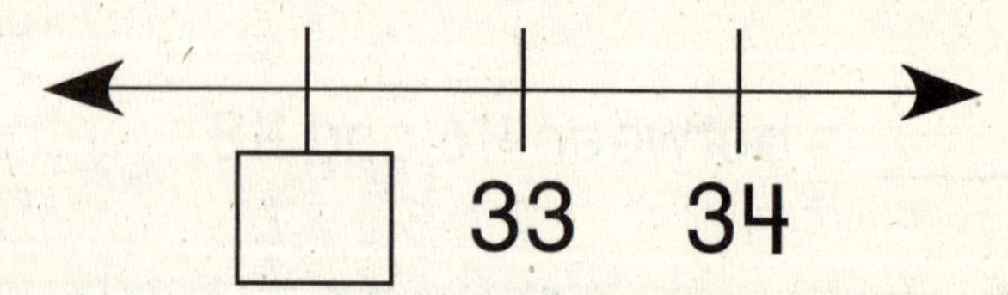

11.

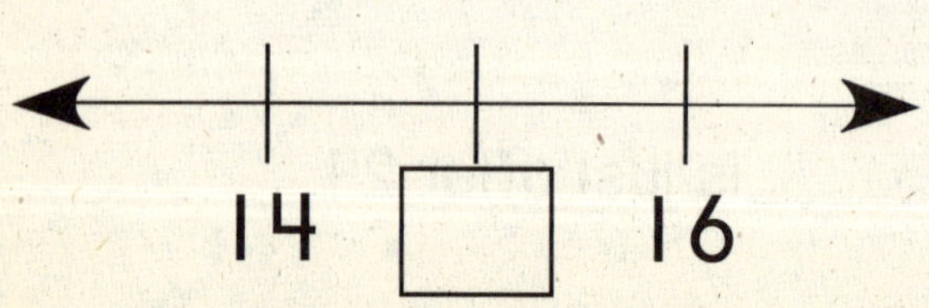

12.

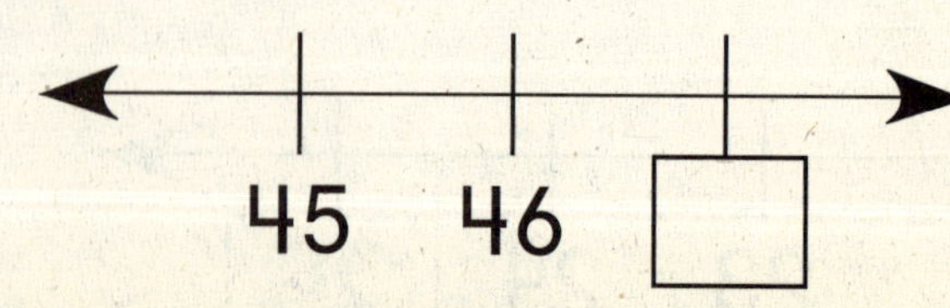

Name______________________

Learn the Math

Add 1. Look for a pattern.

Vocabulary

pattern

1 + 1 = 2

2 + 1 = 3

3 + 1 = 4

Add 2. Look for a pattern.

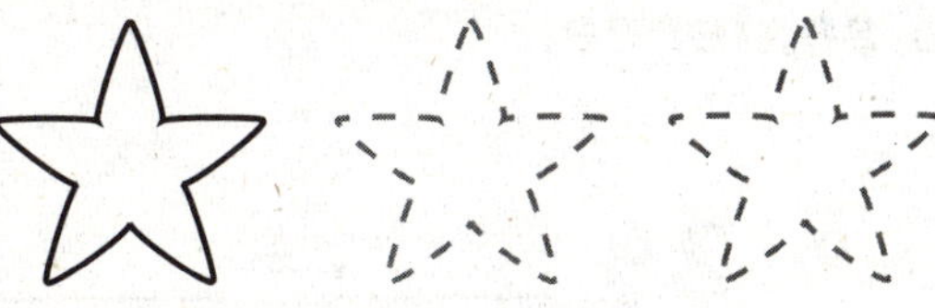

1 + 2 = 3

   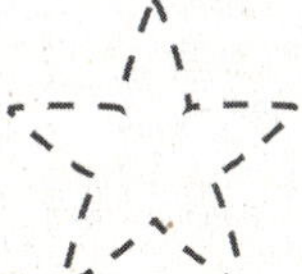

2 + 2 = ___

   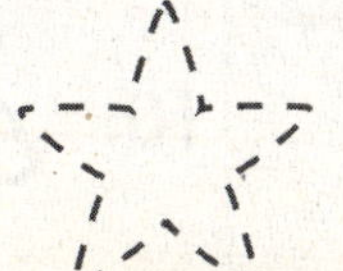 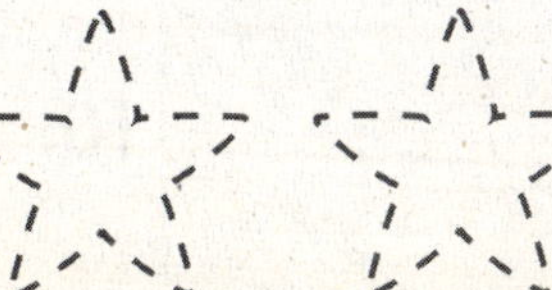 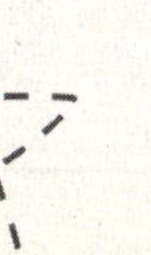

3 + 2 = ___

Do the Math

Skill 9

Count how many. Draw one more.
Write how many in all.

1.  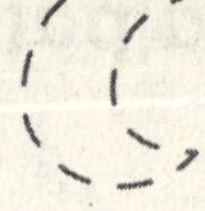

3 +  = 4

2. 

4 + 1 = ____

3. 

5 + 1 = ____

4. 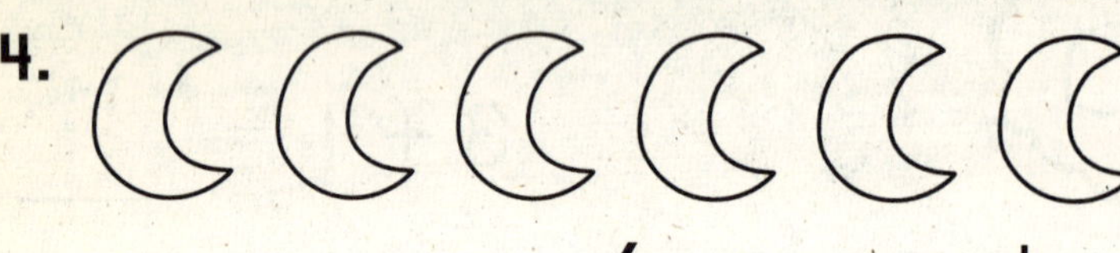

6 + 1 = ____

Add 2. Complete each addition sentence.

5. 1 + 2 = ____

6. 2 + ____ = ____

7. 3 + ____ = ____

8. 4 + ____ = ____

9. 5 + ____ = ____

10. 6 + ____ = ____

Name______________________________________

Subtraction Patterns

Skill 10

Learn the Math

Subtract 1. Look for a pattern.

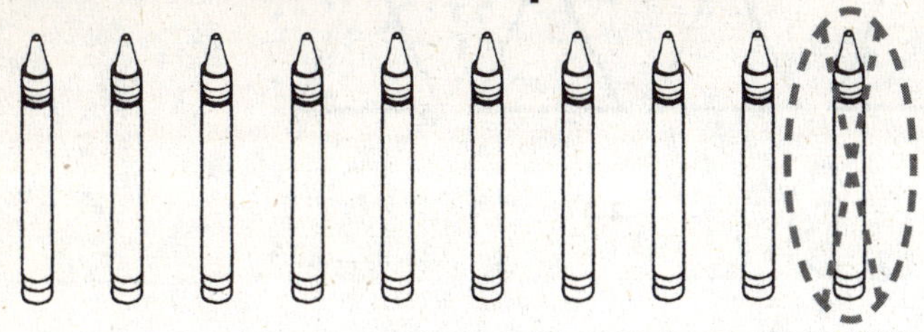

10 − 1 = 9

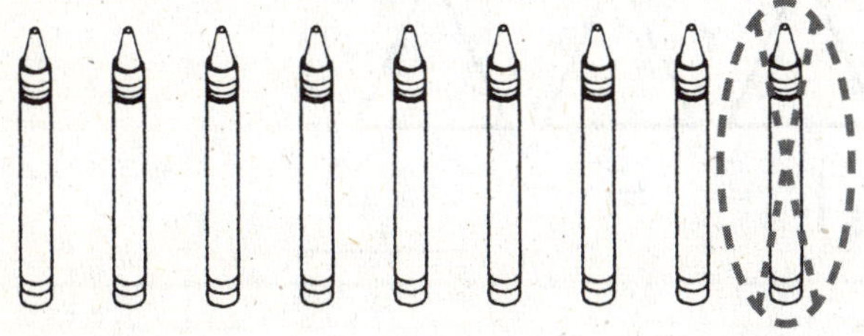

9 − 1 = 8

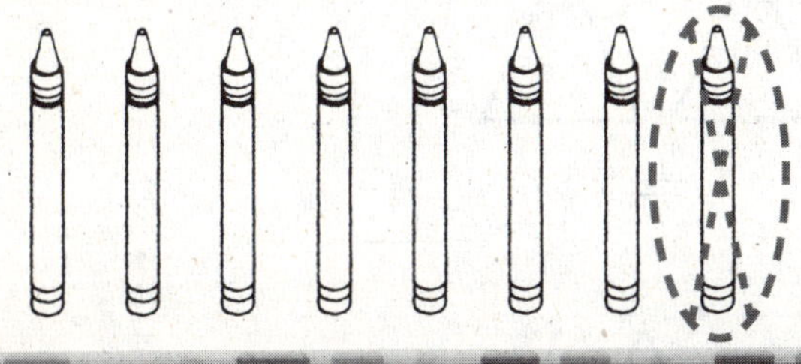

8 − 1 = 7

Subtract 2. Look for a pattern.

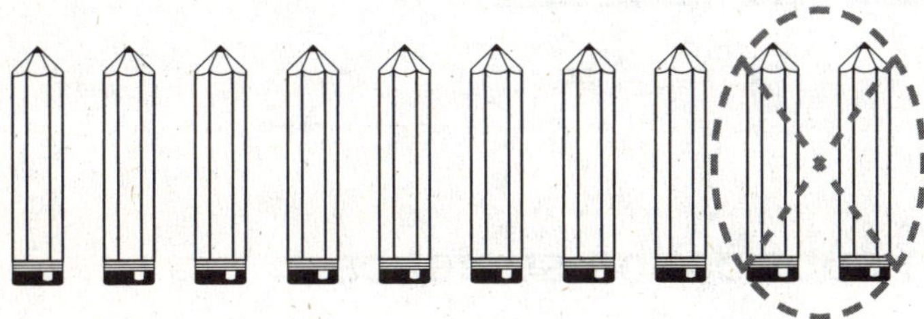

10 − 2 = 8

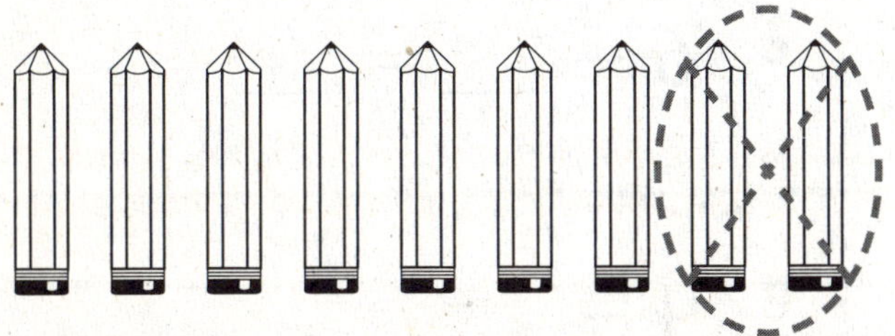

9 − 2 = ____

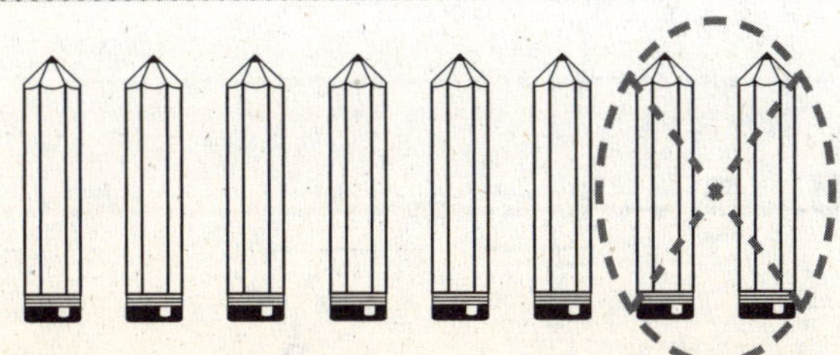

8 − 2 = ____

Circle and mark an X to subtract.
Complete the subtraction sentence.

1. 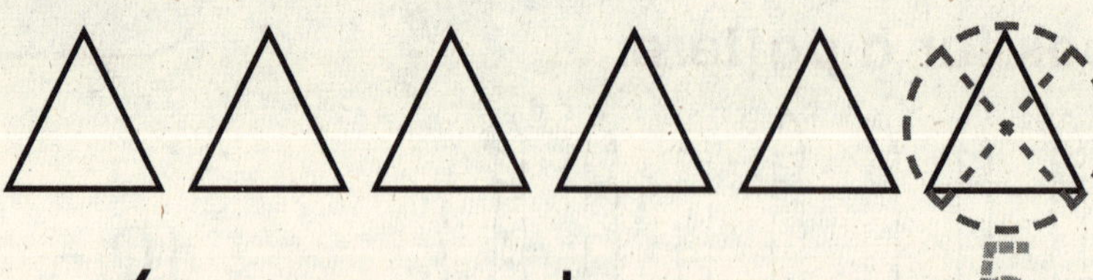

6 − 1 = 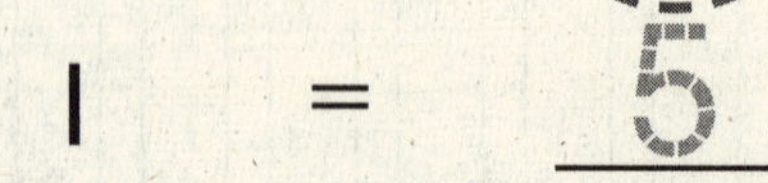

2. 

5 − 1 = ___

3. 

4 − 1 = ___

4. 

3 − 1 = ___

Subtract 2. Complete each subtraction sentence.

5. 7 − 2 = ___

6. 6 − ___ = ___

7. 5 − ___ = ___

8. 4 − ___ = ___

9. 3 − ___ = ___

10. 2 − ___ = ___

Name________________________________

Use a Number Line to Count On

Skill 11

count on

addend

Learn the Math

You can use a number line to count on.

$2 + 6 = \underline{?}$

Which addend is greater? <u>6</u>

So, start on 6.

Then move 2 spaces to the right.

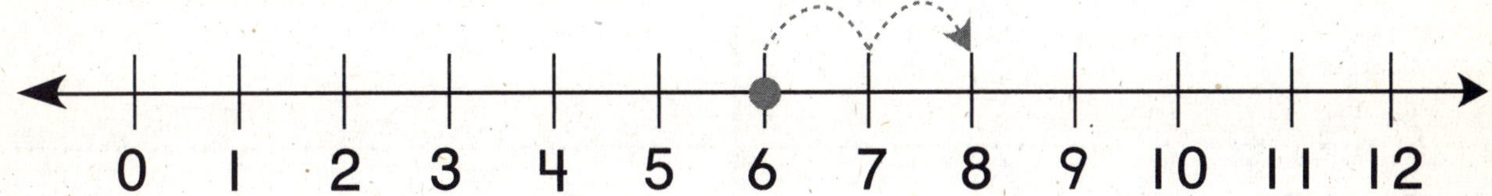

So, $2 + 6 =$ <u>8</u>.

Use the number line to count on.

$4 + 3 = \underline{?}$

Start on <u>4</u>, the greater addend.

Move 3 spaces to the right.

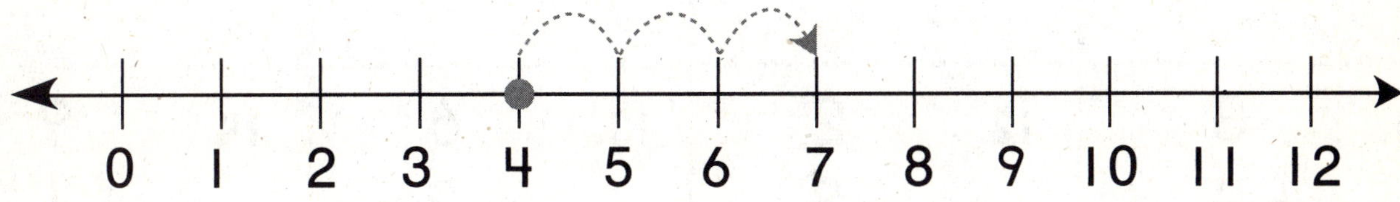

So, $4 + 3 =$ ______.

Do the Math

Use the number line to count on.
Write the sum.

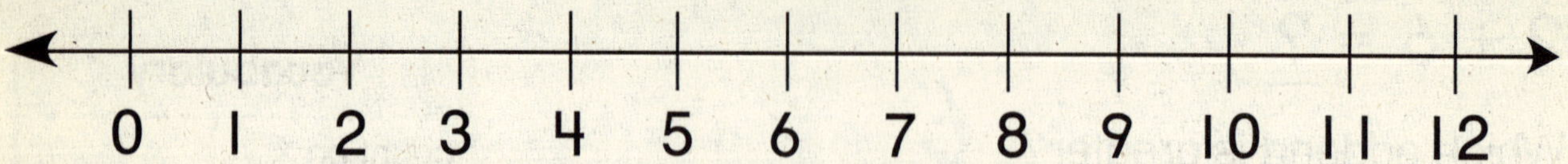

1. $9 + 2 =$ 11

2. $3 + 1 =$ ___

3. $8 + 3 =$ ___

4. $3 + 6 =$ ___

5. $2 + 1 =$ ___

6. $7 + 2 =$ ___

7. $7 + 3 =$ ___

8. $6 + 1 =$ ___

9. $1 + 8 =$ ___

10. $9 + 3 =$ ___

11. $3 + 4 =$ ___

12. $2 + 8 =$ ___

13. $6 + 2 =$ ___

14. $9 + 1 =$ ___

Name ______________________________

Add Doubles
Skill 12

Learn the Math

You can add doubles.
A doubles fact has two addends that are the same.

Vocabulary

doubles fact
addend

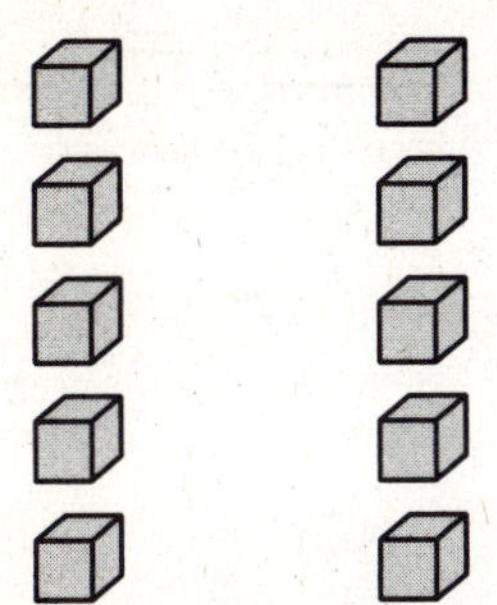

5 + 5 = 10

Write the doubles fact.

4 + 4 = 8

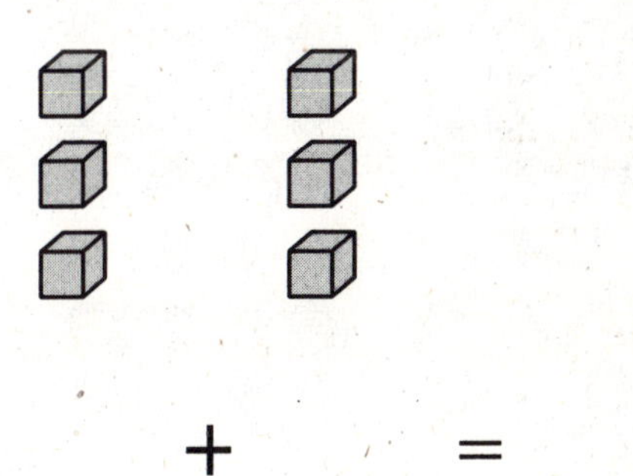

___ + ___ = ___

___ + ___ = ___

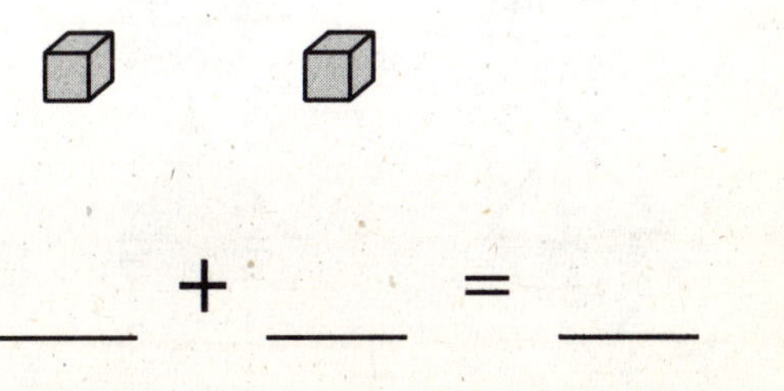

___ + ___ = ___

Do the Math

Write the doubles fact.

1.

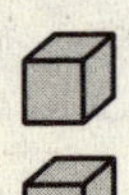

3 + 3 = 6

2.

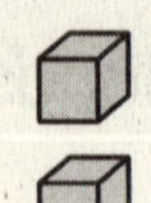

___ + ___ = ___

3.

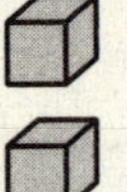

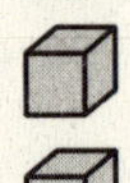

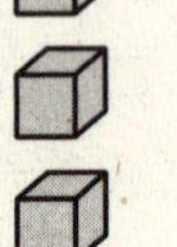

___ + ___ = ___

4.

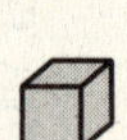

___ + ___ = ___

Write the doubles fact for each picture.

5.

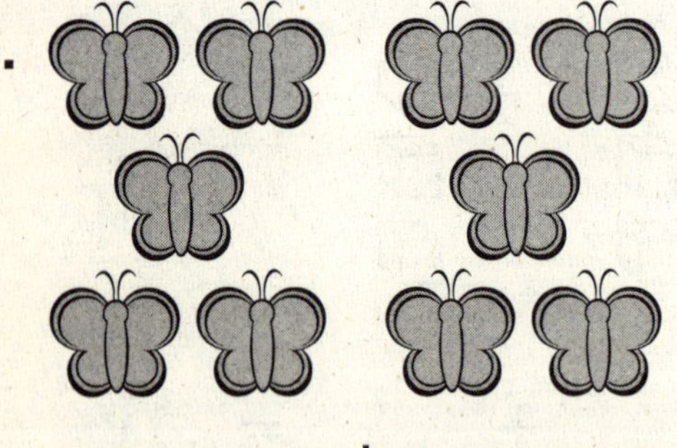

___ + ___ = ___

6.

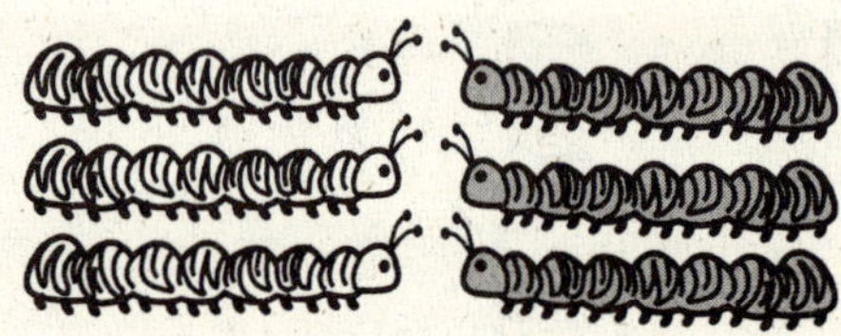

___ + ___ = ___

7.

___ + ___ = ___

8.

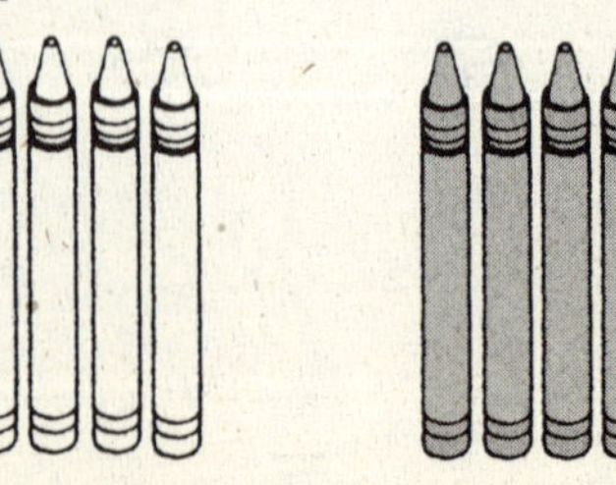

___ + ___ = ___

Name ______________________________

Doubles Plus One

Skill 13

Learn the Math

You can use a doubles-plus-one fact to find a sum.

Vocabulary

doubles plus one

3 + 3 = 6

3 + 3
The addends are the same.
It is a doubles fact.

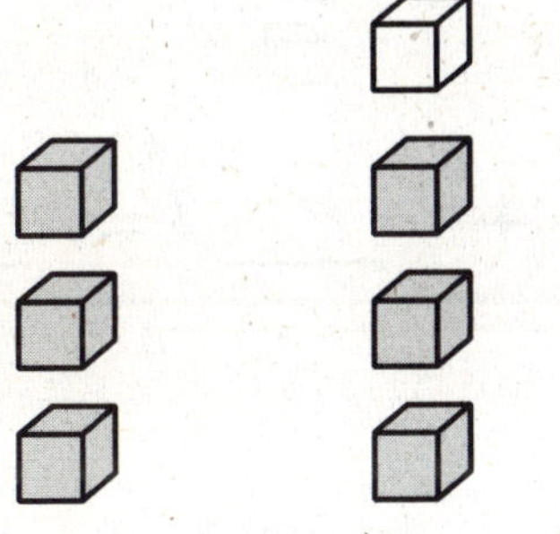

3 + 4 = 7

3 + 4
4 is one more than 3.
It is a doubles-plus-one fact.
Think: 3 + 3 and 1 more.

Find the sum.

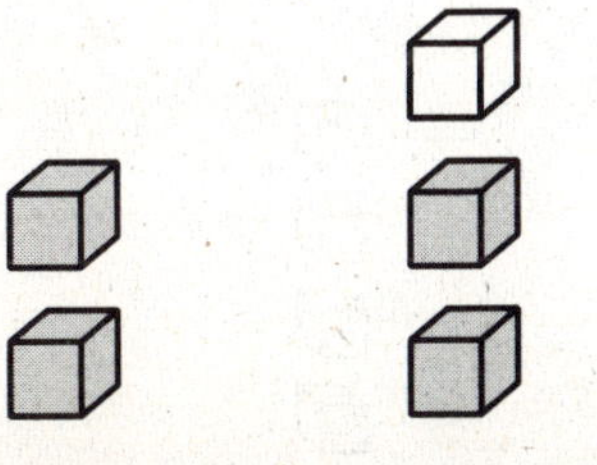

2 + 3 = ____

2 + 3
3 is one more than 2.
It is a doubles-plus-one fact.
Think: 2 + 2 and 1 more.

Do the Math

Use 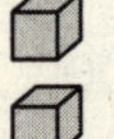 . Write the addition sentence.

1.

4 + 5 = 9

2.

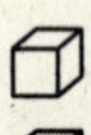

____ + ____ = ____

3.

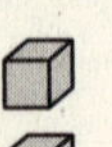
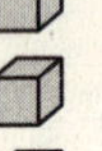

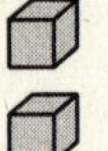

____ + ____ = ____

4.

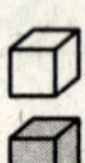
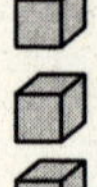
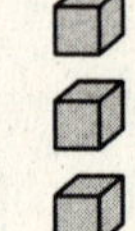

____ + ____ = ____

5.

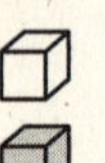
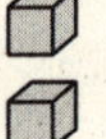
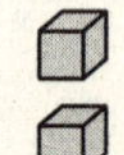

____ + ____ = ____

6.

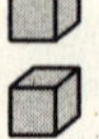

____ + ____ = ____

7.

____ + ____ = ____

8.

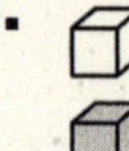
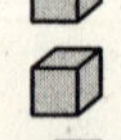
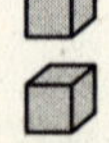
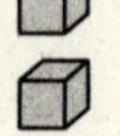
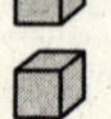
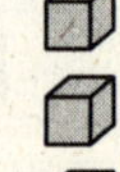

____ + ____ = ____

Name ______________________________

Add in Any Order

Skill 14

Learn the Math

You can add in any order.
The sum is the same.

Vocabulary

addend
order

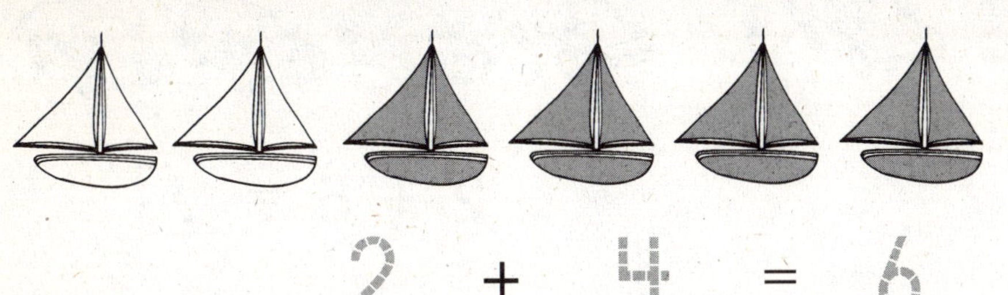

2 + 4 = 6

addend, addend, sum

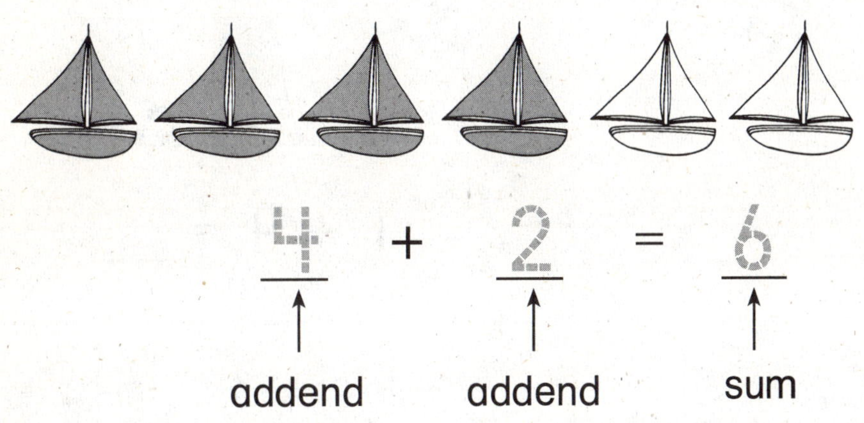

Write the sum.

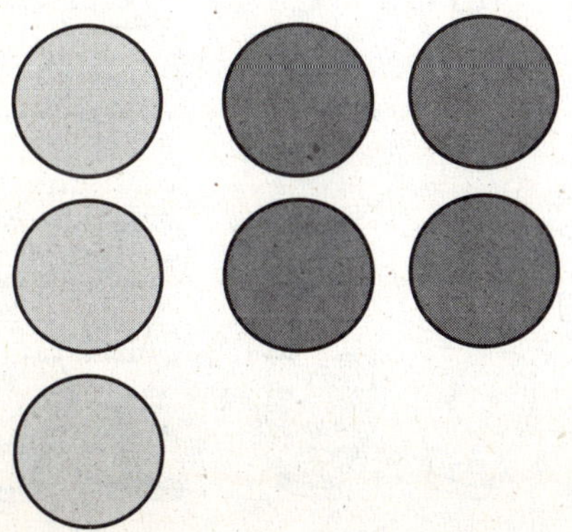

3 + 4 = ___

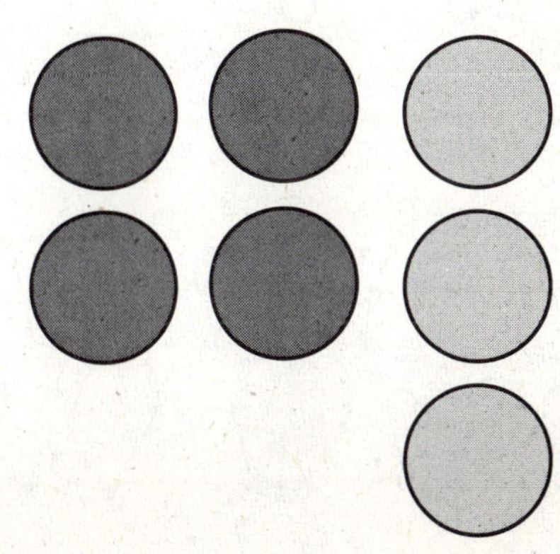

4 + 3 = ___

Write the sum.

1.

$5 + 2 =$ 7 $\qquad 2 + 5 =$ 7

2.

$4 + 1 =$ ___ $\qquad 1 + 4 =$ ___

3.

$3 + 6 =$ ___ $\qquad 6 + 3 =$ ___

4.

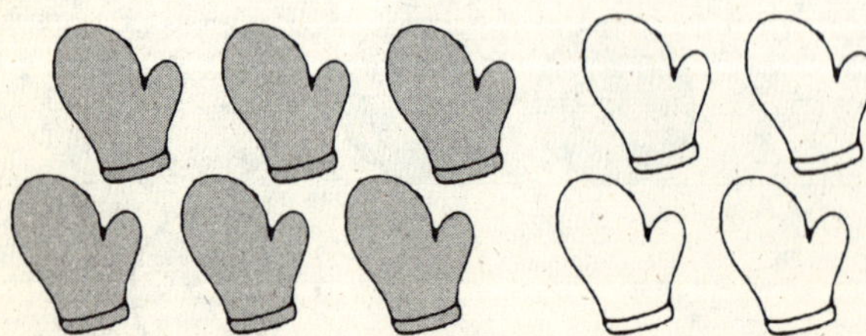

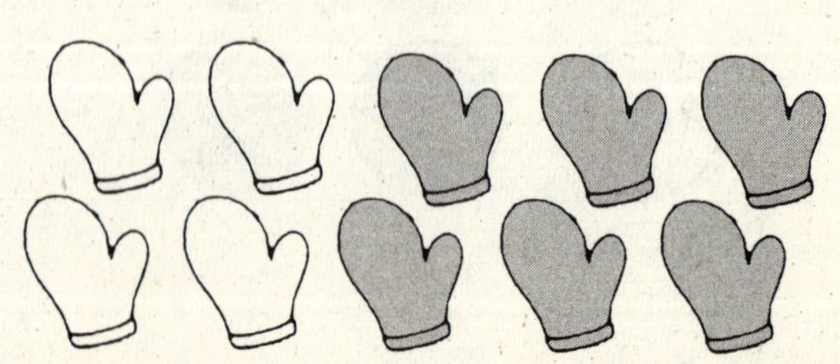

$6 + 4 =$ ___ $\qquad 4 + 6 =$ ___

Name_______________________________

Model Adding Tens

Skill 15

Learn the Math

You can use base-ten blocks to add tens.

40 + 20 = __?__

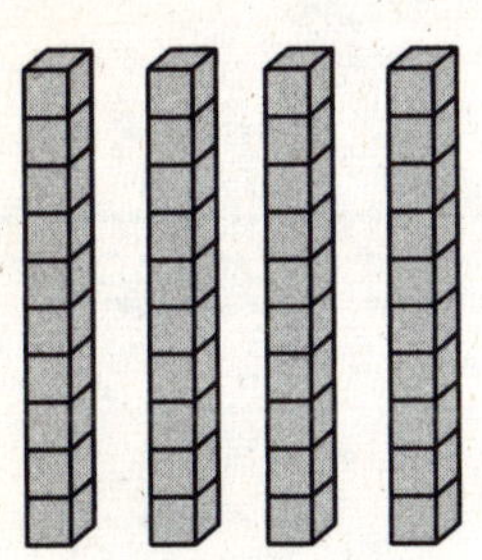

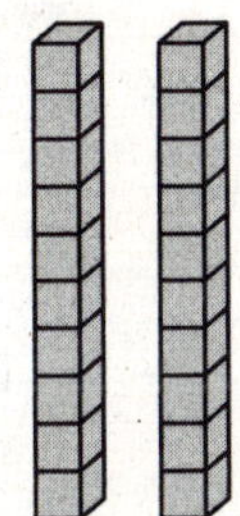

4 tens + 2 tens = __6__ tens

40 + 20 = __60__

---

30 + 50 = __?__

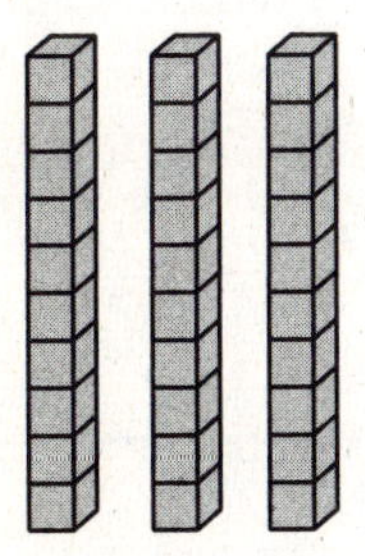

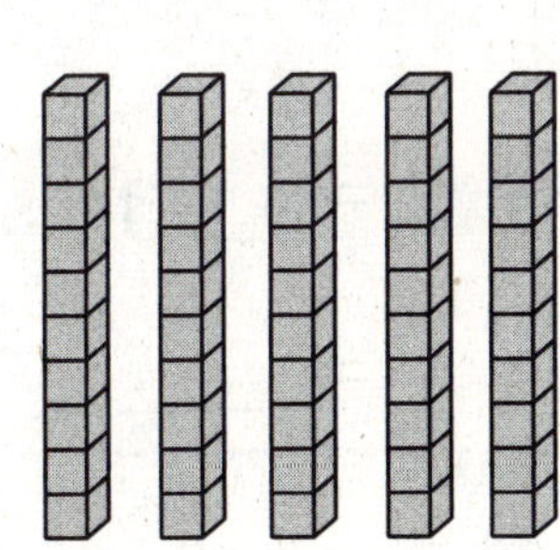

3 tens + 5 tens = ___ tens

30 + 50 = ___

Use base-ten blocks. Find each sum.

1\. 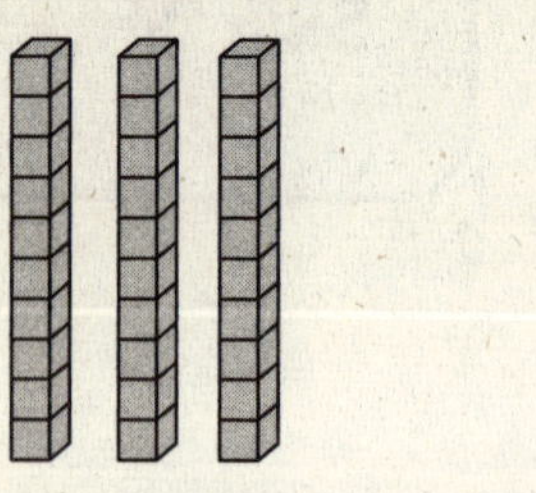 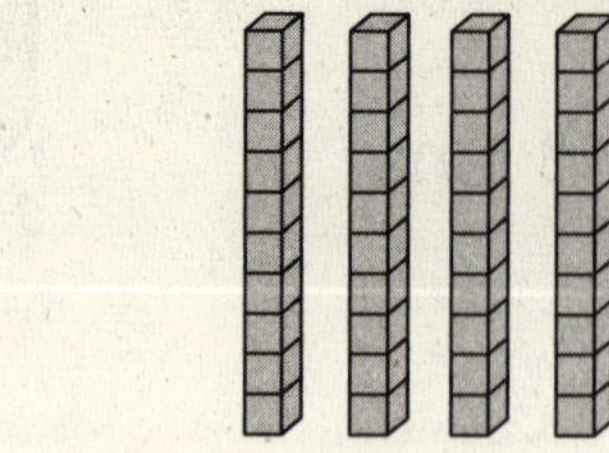

3 tens + 4 tens = 7 tens

30 + 40 = 70

2\. 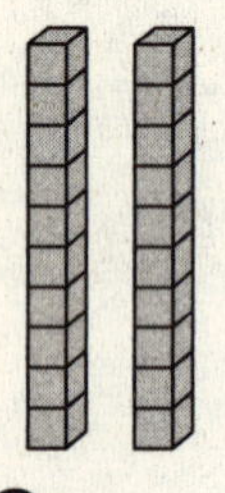 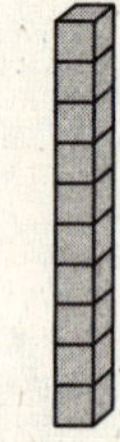

2 tens + 1 ten = ___ tens

20 + 10 = ___

3\. 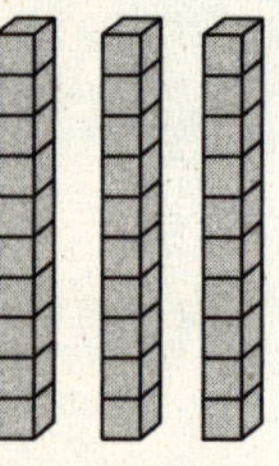 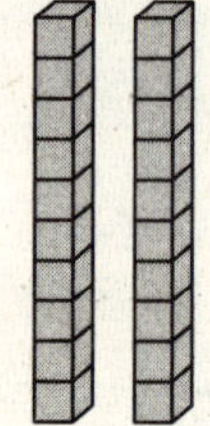

3 tens + 2 tens = ___ tens

30 + 20 = ___

4\. 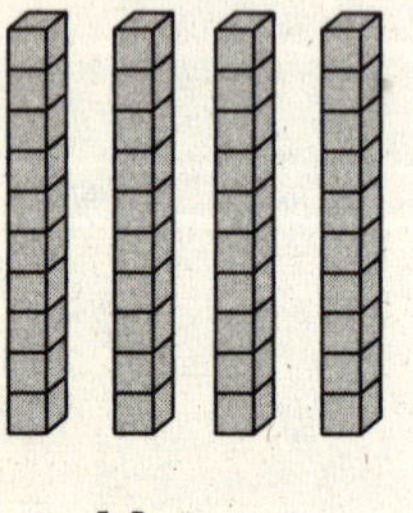 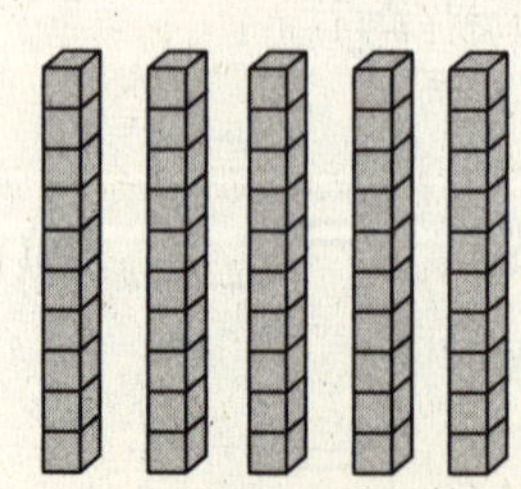

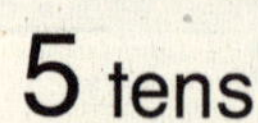

4 tens + 5 tens = ___ tens

40 + 50 = ___

Name____________________

Use a Number Line to Count Back

Skill 16

Learn the Math

You can count back on a number line to subtract.

Vocabulary

count back
number line

$9 - 2 = \underline{?}$

Step 1 Put your finger on 9.

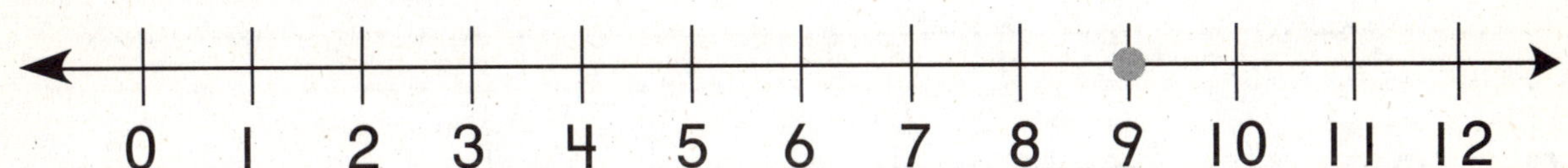

Step 2 Move 2 spaces to the left.

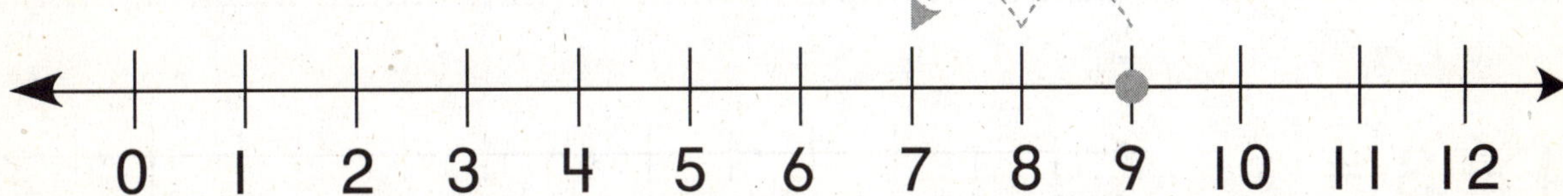

On what number do you stop? 7

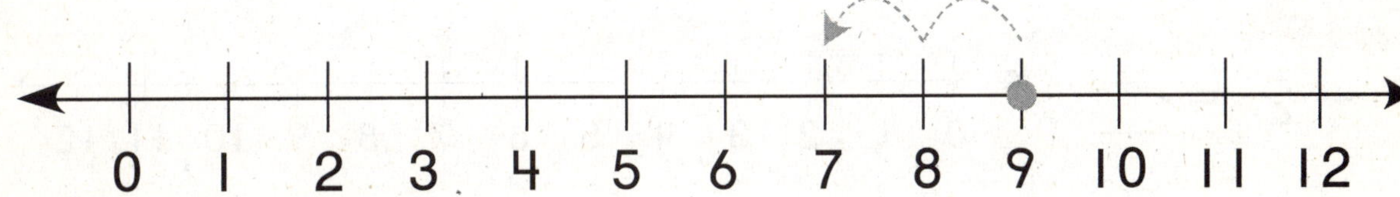

So, $9 - 2 = \underline{7}$.

Use the number line. Count back to subtract.

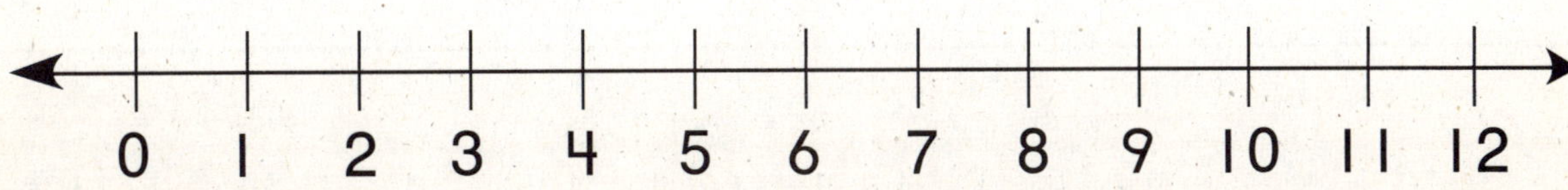

$7 - 3 = \underline{\quad}$

Do the Math

Skill 16

Use the number line. Count back to subtract.

1. 6 − 2 = 4

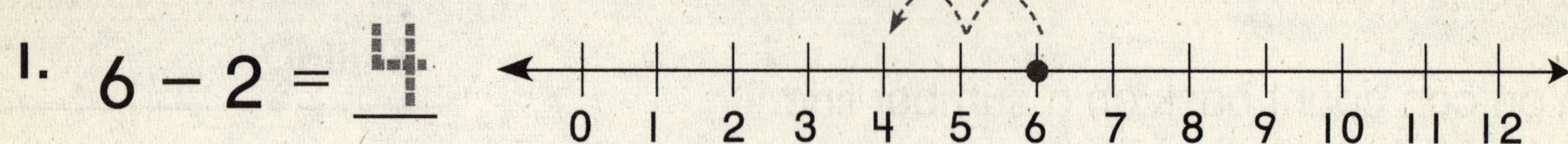

2. 10 − 1 = ___

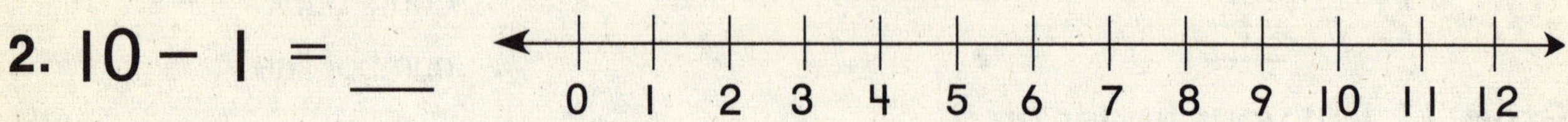

3. 5 − 2 = ___

0 1 2 3 4 5 6 7 8 9 10 11 12

4. 11 − 3 = ___

0 1 2 3 4 5 6 7 8 9 10 11 12

5. 7 − 2 = ___

0 1 2 3 4 5 6 7 8 9 10 11 12

6. 8 − 1 = ___

0 1 2 3 4 5 6 7 8 9 10 11 12

7. 9 − 3 = ___

0 1 2 3 4 5 6 7 8 9 10 11 12

Name________________________________

Related Subtraction Facts to 12

Skill 17

Learn the Math

Related subtraction facts use the same three numbers.

Vocabulary

related subtraction facts

$$\begin{array}{r} 10 \\ -\ 3 \\ \hline 7 \end{array}$$

What numbers are used? 3, 7, 10

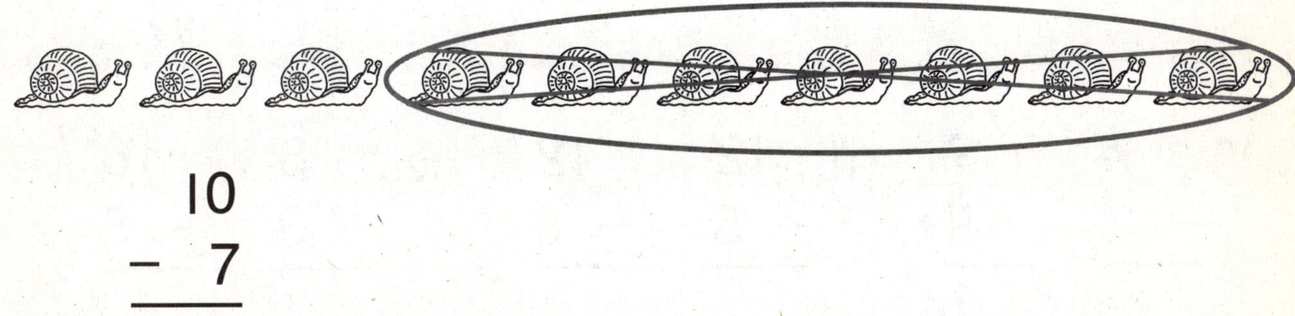

$$\begin{array}{r} 10 \\ -\ 7 \\ \hline \end{array}$$

What numbers are used? ___, ___, ___

So, $10 - 3 = 7$ and $10 - 7 = 3$ are related subtraction facts.

Do the Math

Skill 17

Subtract. Circle the pairs of related subtraction facts.

1. $9 - 6 = 3$ $\quad$ $9 - 3 = 6$

2. $8 - 2 = $ ___ $\quad$ $8 - 6 = $ ___

3. $10 - 4 = $ ___ $\quad$ $10 - 5 = $ ___

4. $12 - 7 = $ ___ $\quad$ $12 - 5 = $ ___

5. $8 - 3 = $ ___ $\quad$ $8 - 4 = $ ___

6. $11 - 6 = $ ___ $\quad$ $11 - 5 = $ ___

7. $12 - 9 = $ ___ $\quad$ $12 - 3 = $ ___

8. $11 - 3 = $ ___ $\quad$ $11 - 8 = $ ___

9. $10 - 9 = $ ___ $\quad$ $10 - 1 = $ ___

10. $9 - 2 = $ ___ $\quad$ $9 - 4 = $ ___

11. $12 - 5 = $ ___ $\quad$ $12 - 6 = $ ___

12. $8 - 3 = $ ___ $\quad$ $8 - 5 = $ ___

13. $12 - 2 = $ ___ $\quad$ $12 - 8 = $ ___

14. $10 - 8 = $ ___ $\quad$ $10 - 2 = $ ___

15. $9 - 7 = $ ___ $\quad$ $9 - 2 = $ ___

Name______________________________________

Fact Families to 12

Skill 18

Learn the Math

A fact family is a group of related addition and subtraction sentences. They all use the same numbers.

Vocabulary

fact family

What is the fact family for 2, 3, and 5?

Step 1 Write related addition sentences.

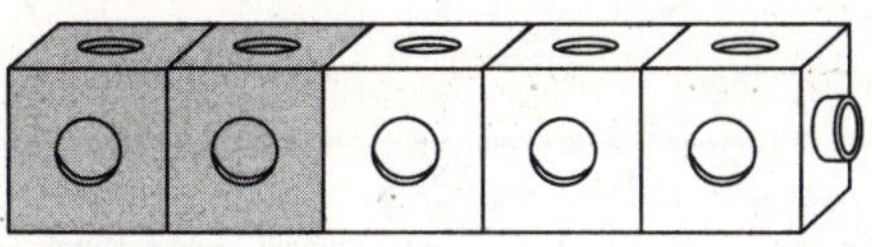

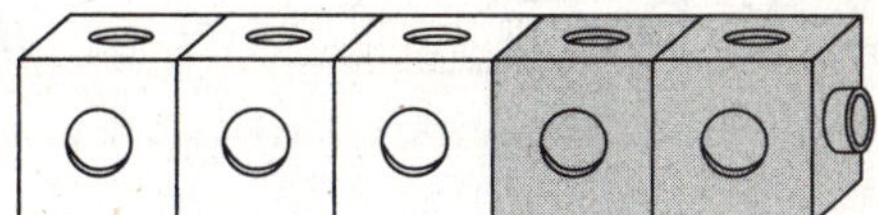

2 + 3 = 5   3 + 2 = 5

Step 2 Write related subtraction sentences.

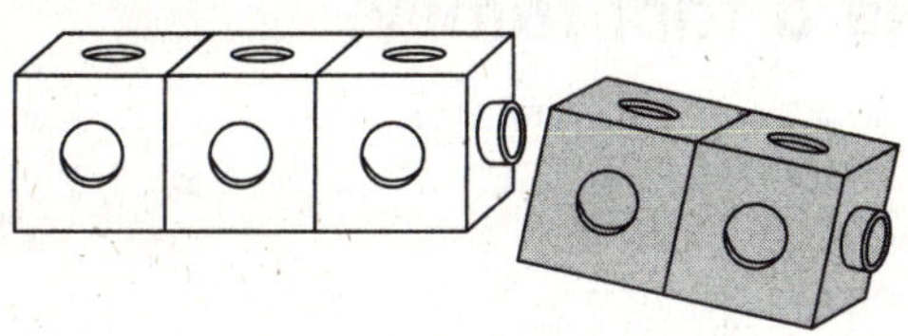

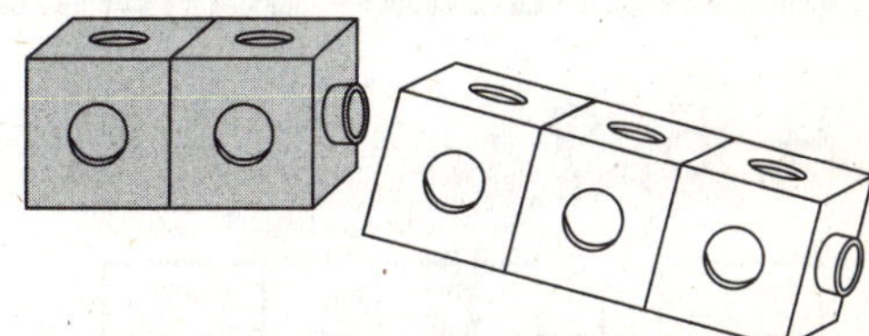

5 − 2 = 3   5 − 3 = 2

So, the fact family for 2, 3, and 5 is:

___ + ___ = ___   ___ − ___ = ___

___ + ___ = ___   ___ − ___ = ___

Add or subtract to complete the fact family.
Write the numbers in the fact family.

1. 6 + 3 = 9   9 − 6 = 3   3, 6, 9

   3 + 6 = 9   9 − 3 = 6

2. 5 + 2 = ___   7 − 5 = ___   ☐, ☐, ☐

   2 + 5 = ___   7 − 2 = ___

3. 1 + 4 = ___   5 − 1 = ___   ☐, ☐, ☐

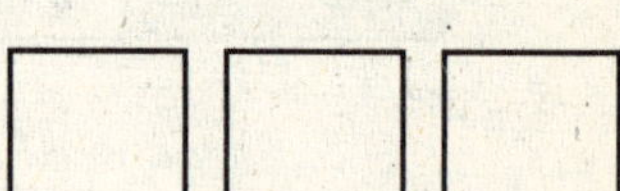

   4 + 1 = ___   5 − 4 = ___

4. 5 + 5 = ___   10 − 5 = ___   ☐, ☐

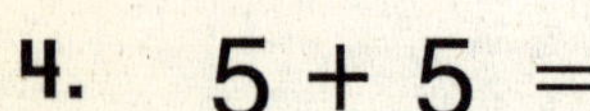

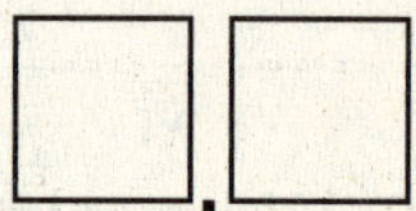

Write the number sentences to make a fact family.

5. 4, 5, 9

   4 + 5 = 9   ☐ − ☐ = ☐

   ☐ + ☐ = ☐   ☐ − ☐ = ☐

6. 3, 7, 10

   ☐ + ☐ = ☐   ☐ − ☐ = ☐

   ☐ + ☐ = ☐   ☐ − ☐ = ☐

Name ______________________________

Think Addition to Subtract

Skill 19

Learn the Math

You can use addition facts to help you subtract.

8 − 6 = ?

Think: 6 + 2 = 8

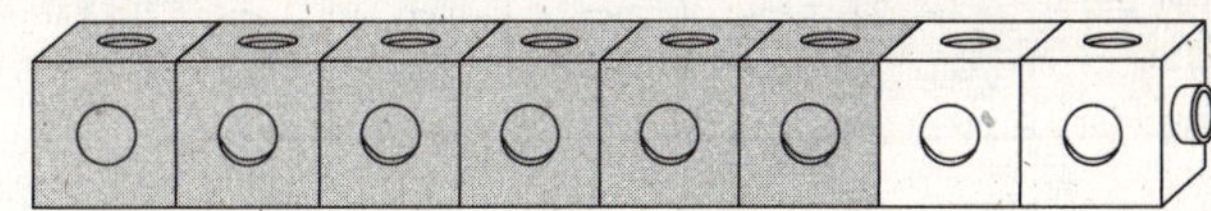

Since 6 + ___ = 8, then 8 − 6 = ___.

---

10 − 3 = ?

Think: 3 + ___ = 10

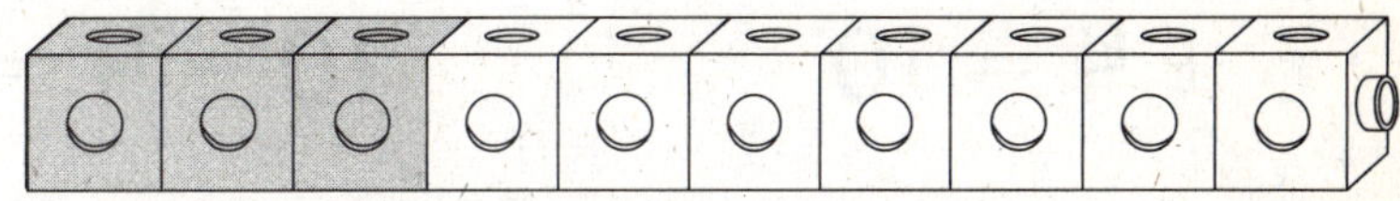

Since 3 + ___ = 10, then 10 − 3 = ___.

---

9 − 6 = ?

Think: 6 + ___ = 9

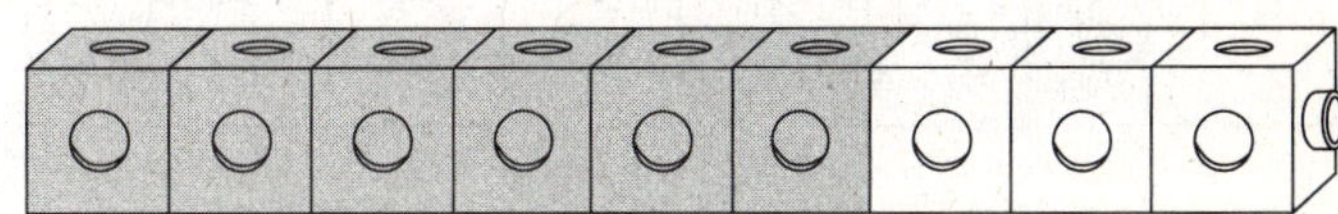

Since 6 + ___ = 9, then 9 − 6 = ___.

Find each difference. Use the addition fact to help you.

1. 3 + 4 = 7   7 − 4 = 3

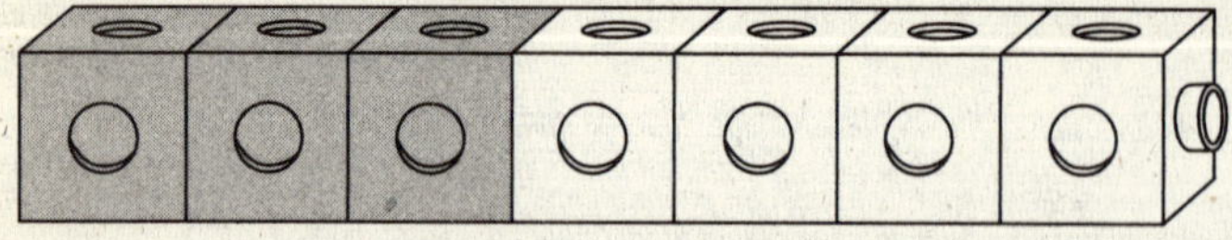

2. 5 + 3 = 8   8 − 3 = ___

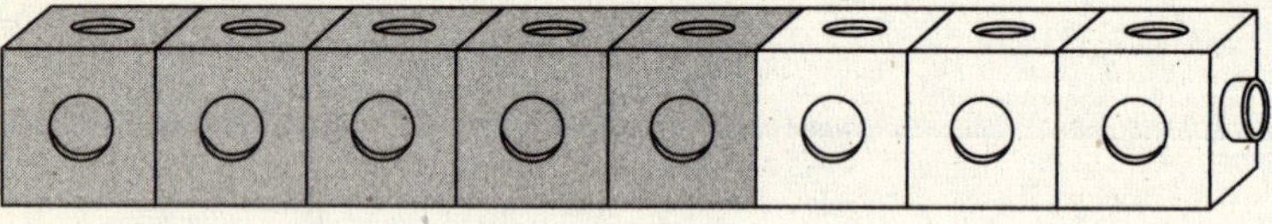

3. 6 + 4 = 10   10 − 4 = ___

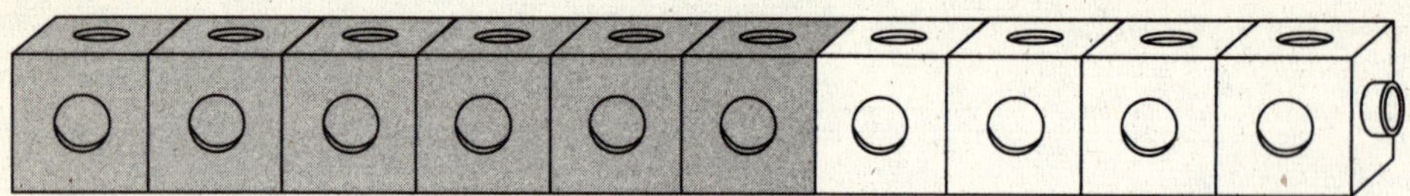

4. 4 + 5 = 9   9 − 5 = ___

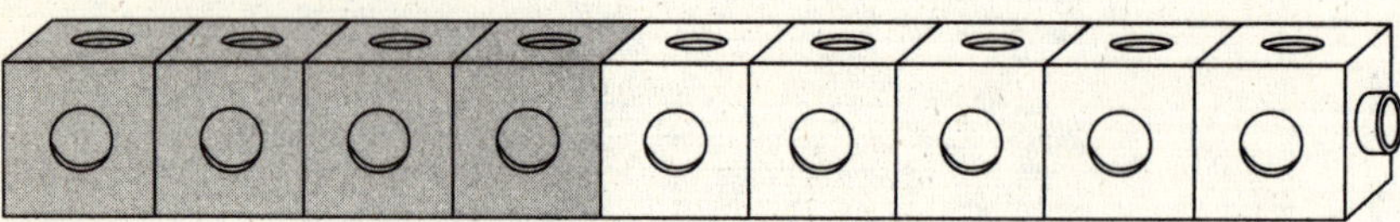

5. 7 + 5 = 12   12 − 5 = ___

Name________________________________

Model Subtracting Tens

Skill 20

Learn the Math

You can use base-ten blocks to subtract tens.

50 − 20 = ?

How many tens are there in all? __5__ tens

How many tens are taken away? __2__ tens

How many tens are left? __3__ tens

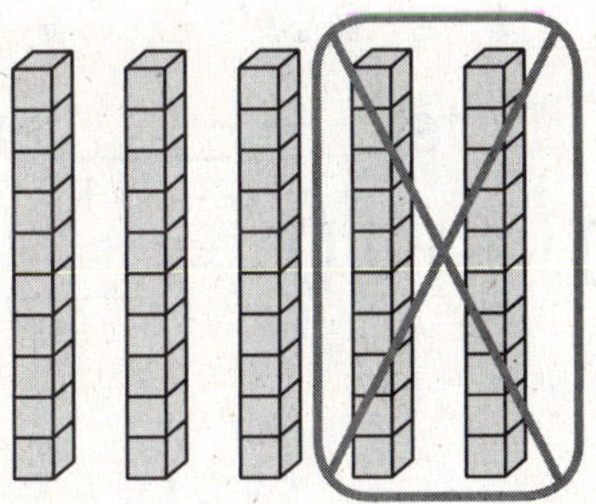

5 tens − 2 tens = __3__ tens

50 − 20 = __30__

---

60 − 40 = ?

How many tens in are there all? ____ tens

How many tens are taken away? ____ tens

How many tens are left? ____ tens

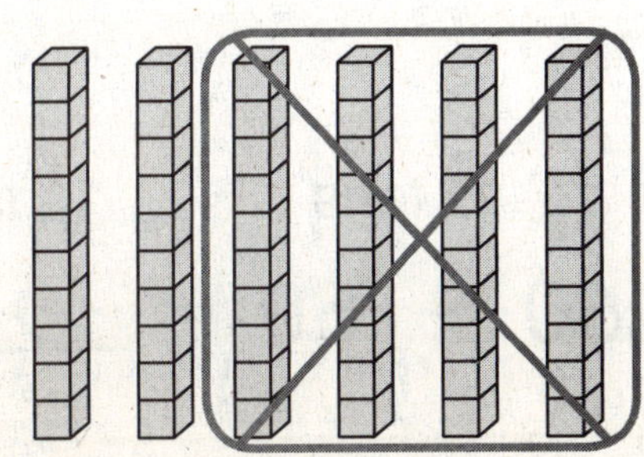

6 tens − 4 tens = ____ tens

60 − 40 = ____

Use base-ten blocks. Find each difference.

1.

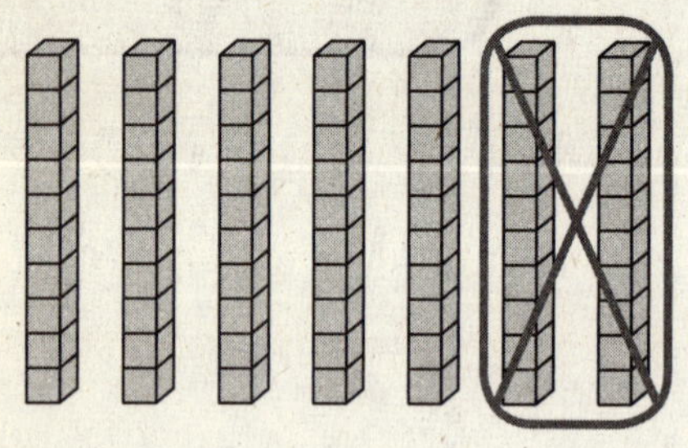

7 tens − 2 tens = 5 tens

70 − 20 = 50

2.

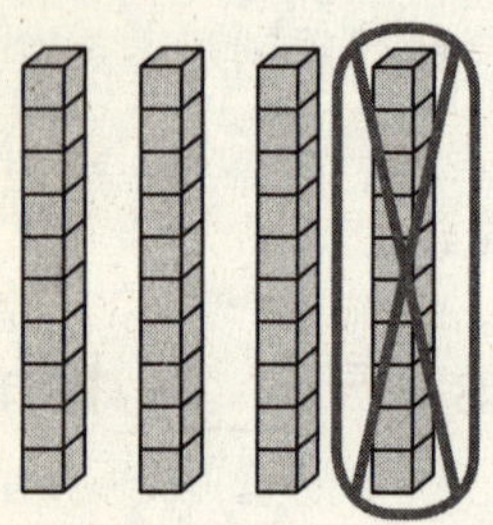

4 tens − 1 ten = ____ tens

40 − 10 = ____

3.

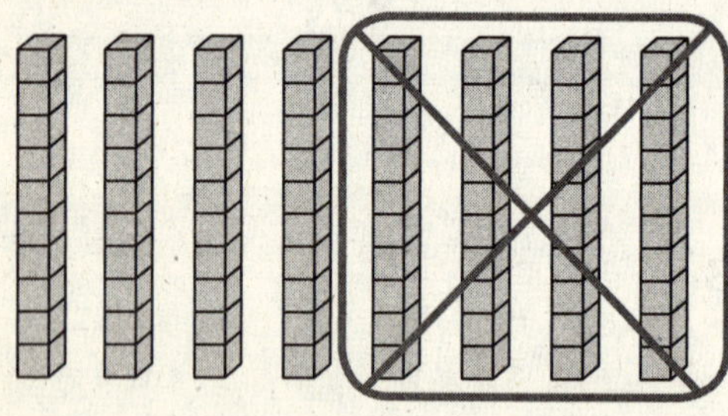

8 tens − 4 tens = ____ tens

80 − 40 = ____

4.

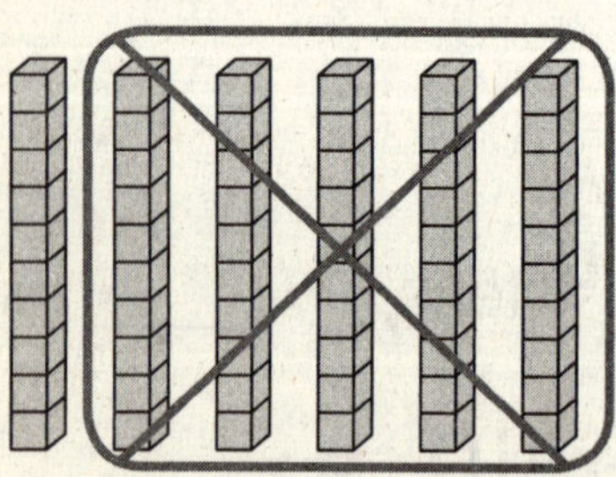

6 tens − 5 tens = ____ ten

60 − 50 = ____

Name________________________________

Read a Tally Table

Skill 21

Learn the Math

Erin has some puppets.

She makes a tally table about her puppets.

Vocabulary

tally table

tally mark

Erin's Puppets	
Puppet	**Tally**
tiger	|||
bear	~~||||~~
monkey	||

Read the title.

What is the tally table about?

Erin's Puppets

Read the label for each row.

What kinds of puppets does Erin have?

tigers, bears, and monkeys

Read the tally marks.

How many of each puppet does Erin have?

 3  ____ ____

Compare.

Of which kind of puppet is there the most? ____________

Use the tally table.

Vegetables in Maya's Garden	
Vegetable	**Tally**
carrot	卌 |||
tomato	卌 |
pepper	|||

1. What is the tally table about?

Vegetables in Maya's Garden

2. How many of each kind are in Maya's garden?

 ____  ____  ____

3. Which of these is there the most?

4. Which of these is there the least?

5. How many vegetables does Maya have in all?

____ vegetables

Name____

Read a Picture Graph

Skill 22

Learn the Math

You can read a picture graph. Use the picture graph to answer the question.

Vocabulary

picture graph

Our Favorite Flowers									
rose									
daisy									
tulip									

How many children chose  **?**

Step 1

Read the question.

How many children chose  ?

Step 2

Count how many .

1, 2, 3, 4

Step 3

Write how many.

____ children

So, ____ children chose  .

Use the picture graph.

Our Favorite Snacks								
apple								
GRANOLA	GRANOLA	GRANOLA	GRANOLA	GRANOLA	GRANOLA	GRANOLA	GRANOLA	
YOGURT	YOGURT	YOGURT	YOGURT	YOGURT	YOGURT			

1. How many children chose ?

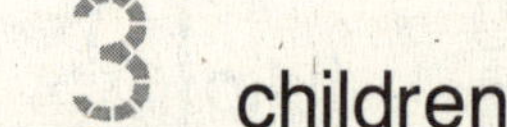

2. How many children chose  ? _____ children

3. How many children chose ? _____ children

4. Circle the snack that most children chose.

Name____

Make a Prediction

Skill 23

Learn the Math

Evie has a bag of tiles.

She has 4 gray tiles and 1 black tile.

Evie pulls a tile.

Which color is Evie more likely to pull?

Vocabulary

more likely

prediction

Step 1

Count how many tiles there are of each color.

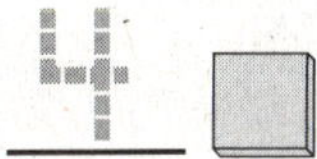

Step 2

Circle the color there is more of.

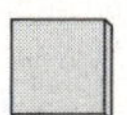

Step 3

Think: There are more gray tiles.
Circle your prediction.

So, Evie is more likely to pull a __________ tile.

Use the pictures to make a prediction.

1. Alex pulls a sock from the bag. Which color sock is Alex more likely to pull?

   Count the socks.

   3 ______

   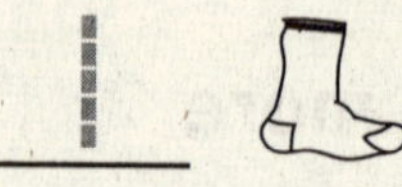

   Alex is more likely to pull  .

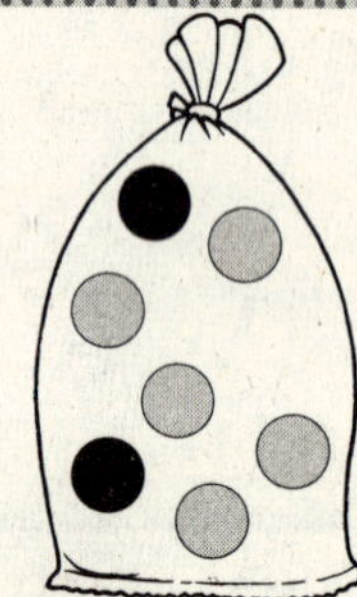

2. Devin pulls a marble from the bag. Which color marble is Devin more likely to pull?

   Count the marbles.

   ______ ______

   Devin is more likely to pull  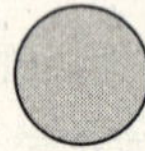.

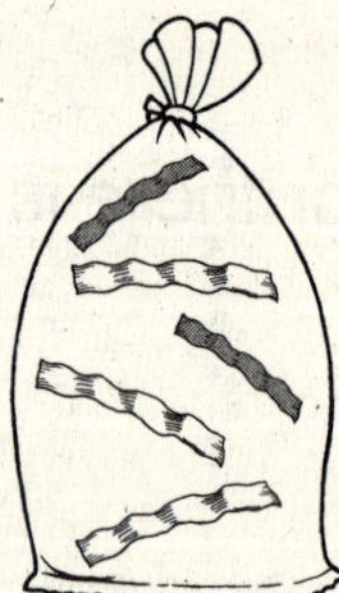

3. Delia pulls a ribbon from the bag. Which color ribbon is Delia more likely to pull?

   Count the ribbons.

   ______ ______

   Delia is more likely to pull  .

Name____

Pennies, Nickels, and Dimes

Skill 24

Learn the Math

You can show the values of pennies, nickels, and dimes.

Vocabulary	
penny	dime
nickel	

Show the value of one dime with pennies.

=

10 cents

1 dime

 pennies

Show the value of one dime with nickels.

=

10 cents

1 dime

____ nickels

Show the value of one nickel with pennies.

=

5 cents

1 nickel

____ pennies

Do the Math

Draw pennies to show equal value.

1.  = 

5 pennies

2.  =

____ pennies

Draw nickels to show equal value.

3.      =

____ nickel

4.  =

____ nickels

Name______________________________

Skip-Count by Fives and Tens

Skill 25

Learn the Math

You can use a picture to skip-count.

How many grapes are there?
Skip-count by fives.

There are 25 grapes.

How many cherries are there?
Skip-count by tens.

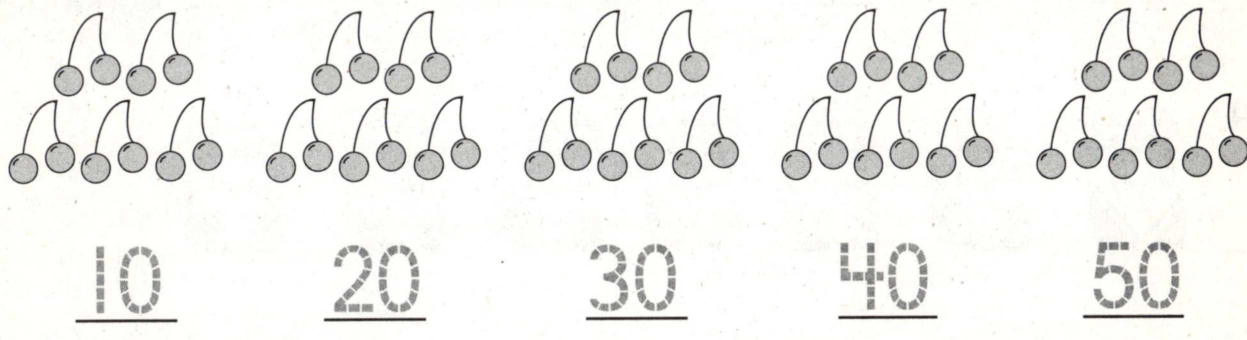

There are _____ cherries.

Do the Math

Skip-count by fives or tens. Write how many.

1.

5 ____ 10 ____ 15 ____ 20 ____ 25 ____ 30 ____ blueberries

2.

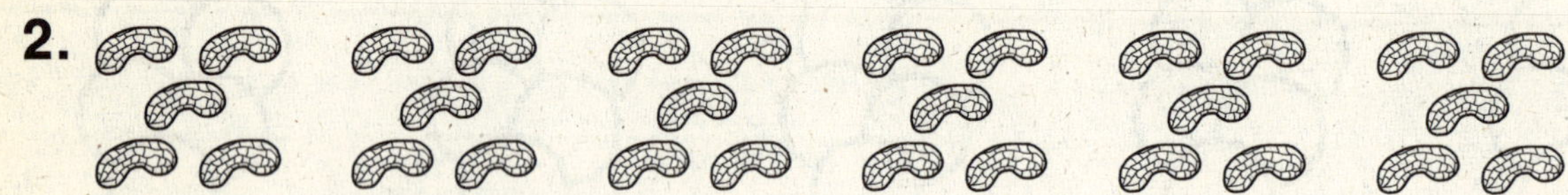

5 ____ ____ ____ ____ ____ ____ peanuts

3.

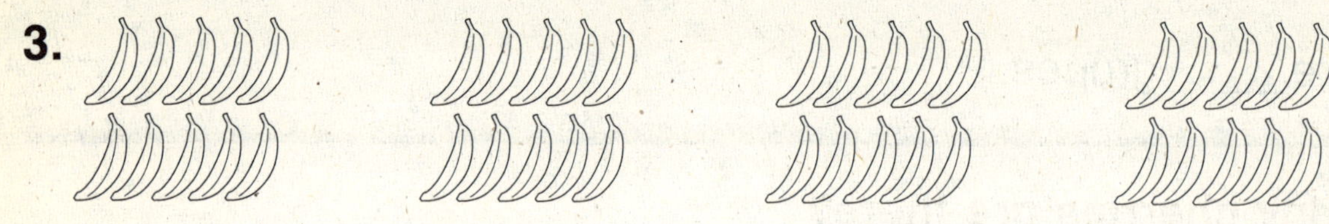

10 ____ ____ ____ ____ bananas

4.

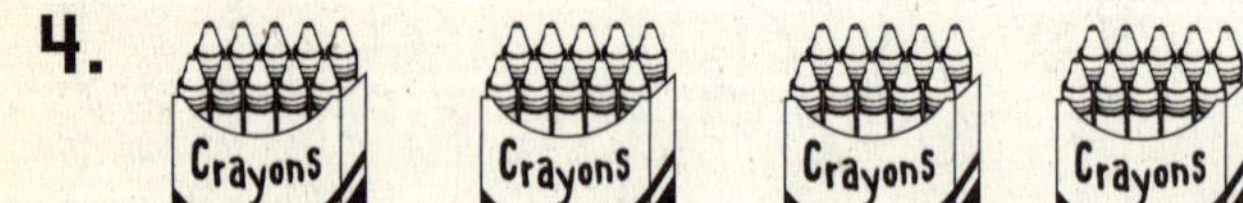

10 ____ ____ ____ ____ ____ ____ ____ crayons

Name________________________________

Count Money

Skill 26

Learn the Math

You can count to find the value of money.

Vocabulary

penny
nickel
dime

Count by ones. Write the total value.

total value

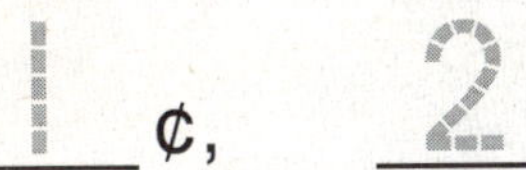 ¢,  ¢  ¢

Count by fives. Write the total value.

_____ ¢, _____ ¢, _____ ¢, _____ ¢ ☐ ¢

Count by tens. Write the total value.

_____ ¢, _____ ¢, _____ ¢, _____ ¢, _____ ¢ ☐ ¢

Count by tens. Count by fives. Then count by ones.
Write the total value.

_____ ¢, _____ ¢, _____ ¢, _____ ¢, _____ ¢, _____ ¢ ☐ ¢

Count. Write the total value.

1.    

total value

1 ¢, 2 ¢, 3 ¢, 4 ¢, 5 ¢, 6 ¢  6 ¢

2.     

____ ¢, ____ ¢, ____ ¢, ____ ¢, ____ ¢ 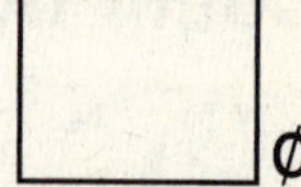 ¢

3.   

____ ¢, ____ ¢, ____ ¢  ¢

4.      

____ ¢, ____ ¢, ____ ¢, ____ ¢, ____ ¢, ____ ¢ 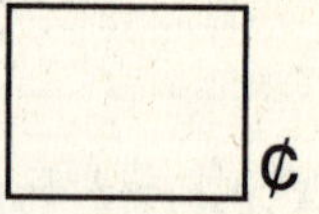 ¢

5.    

____ ¢, ____ ¢, ____ ¢, ____ ¢, ____ ¢, ____ ¢  ¢

Name____

Pennies, Nickels, Dimes, and Quarters

Skill 27

Vocabulary

quarter

Learn the Math

The value of one quarter is 25 cents.

or

1 quarter = 25 cents or 25¢

Count. Write the total value.
Start with the quarter. Count dimes by tens.
Count nickels by fives. Count pennies by ones.

total value

25 ¢, 35 ¢, 40 ¢, 45 ¢,  ¢  ¢

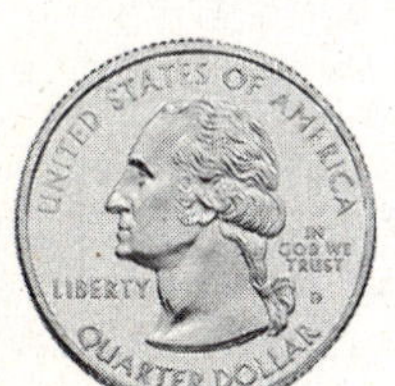     

______ ¢, ______ ¢, ______ ¢, ______ ¢, ______ ¢, ______ ¢ [ ] ¢

Count. Write the total value.

1.     

total value

25 ¢, 35 ¢, 45 ¢, 55 ¢, 56 ¢, 57 ¢ [57] ¢

2.     

____ ¢, ____ ¢, ____ ¢, ____ ¢, ____ ¢ [ ] ¢

3.     

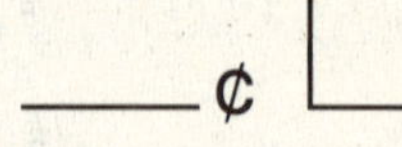

____ ¢, ____ ¢, ____ ¢, ____ ¢, ____ ¢ [ ] ¢

4.      

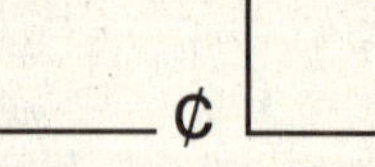

____ ¢, ____ ¢, ____ ¢, ____ ¢, ____ ¢, ____ ¢ [ ] ¢

5.     

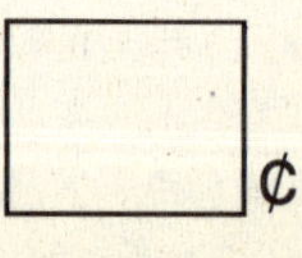

____ ¢, ____ ¢, ____ ¢, ____ ¢, ____ ¢ [ ] ¢

Name______________________________

Use a Clock

Skill 28

Learn the Math

You can use a clock to tell time.

Vocabulary

clock
minute hand
hour hand

Step 1
Look at the minute hand.

minute hand

It points to 12.

Step 2
Look at the hour hand.

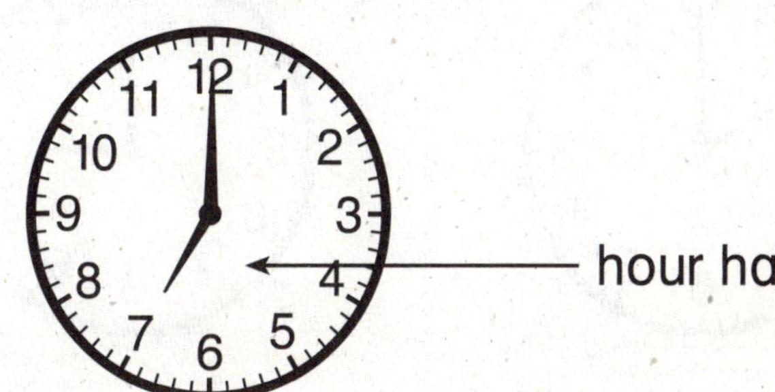

hour hand

It points to 7.

Step 3
Write the time.

7 o'clock

Write the time.

9 o'clock

____ o'clock

____ o'clock

Do the Math

Write the time.

1\.

2 o'clock

2\.

____ o'clock

3\.

____ o'clock

4\.

____ o'clock

5\.

____ o'clock

6\.

____ o'clock

7\.

____ o'clock

8\.

____ o'clock

9\.

____ o'clock

Name________________________________

Minutes and Hours

Skill 29

Learn the Math

You can estimate about how long it takes to do something.

Vocabulary

minute estimate hour

tie your shoes

about 1 minute

cook dinner

about 1 hour

About how long does it take? Circle your answer.

exercise in gym class

about one minute

about one hour

brush your teeth

about two minutes

about two hours

About how long does it take? Circle your answer.

1. write 1 to 10

about 1 minute

about 1 hour

2. watch a play

about 2 minutes

about 2 hours

3. make a sandwich

about 5 minutes

about 5 hours

4. read a story

about 10 minutes

about 10 hours

5. sing a song

about 3 minutes

about 3 hours

6. take a piano lesson

about 1 minute

about 1 hour

Name____

Sort by Color, Size, and Shape

Skill 30

Learn the Math

You can sort objects by color.

All of these are white.

You can sort objects by shape.

All of these are circles.

You can sort objects by size.

All of these are the same size.

Mark an X on the object that does not belong.

Mark an X on the one that does not belong.

1.

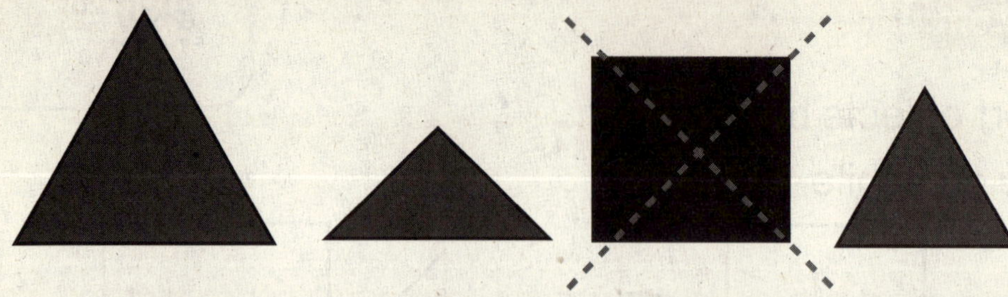

2.

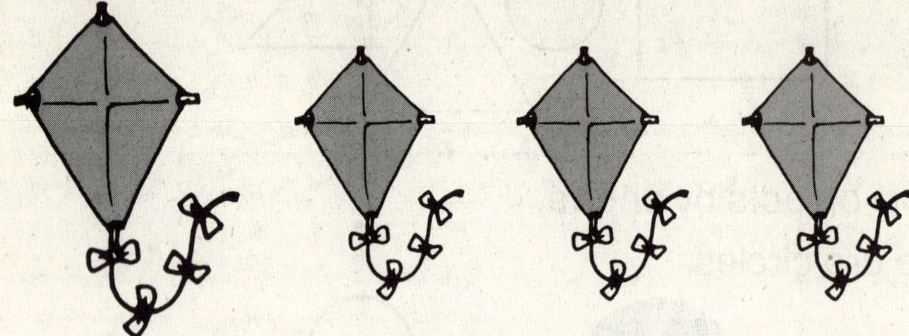

Circle the group in which the object belongs.

3.

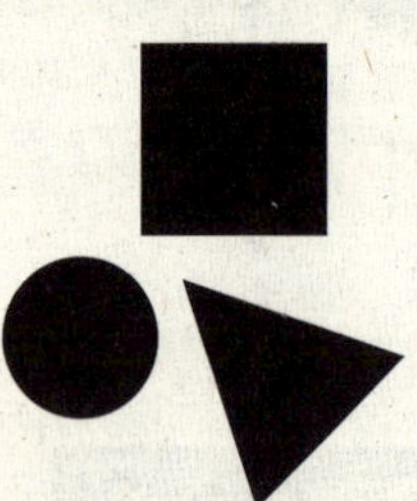

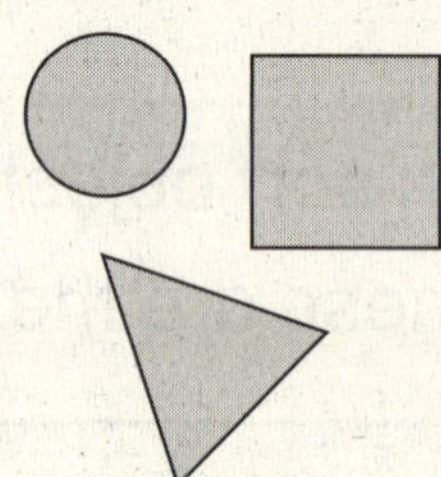

4.

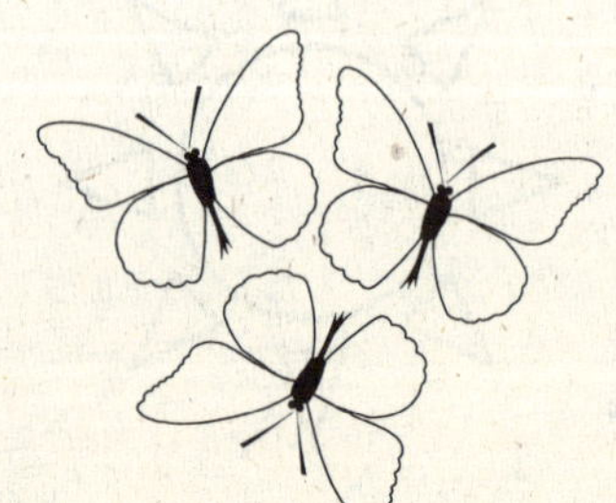

Name____

Identify Three-Dimensional Shapes

Skill 31

Learn the Math

These are three-dimensional shapes.

Vocabulary

cone
cube
cylinder
pyramid
rectangular prism
sphere
three-dimensional

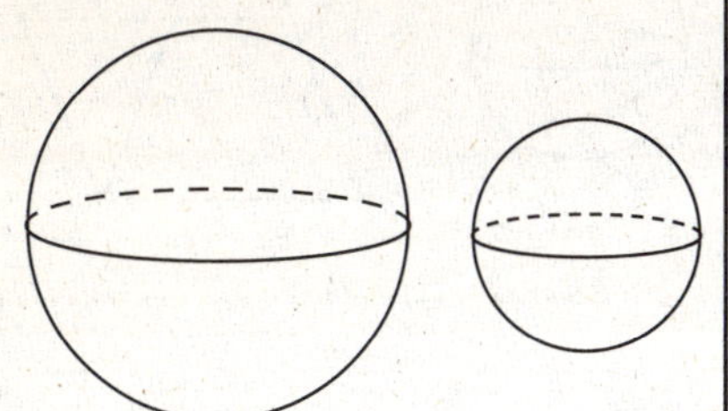

sphere

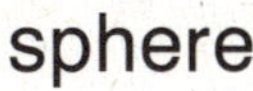

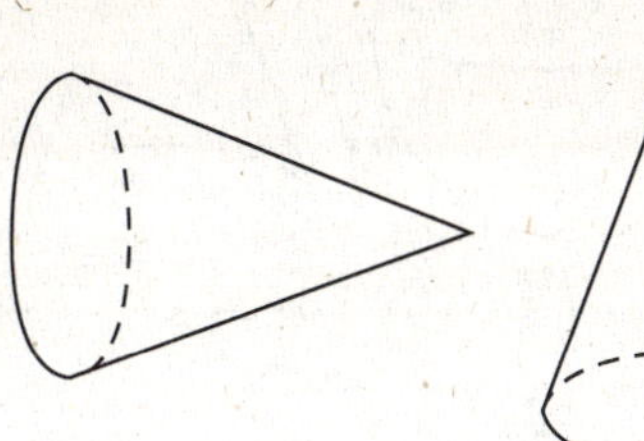

cone

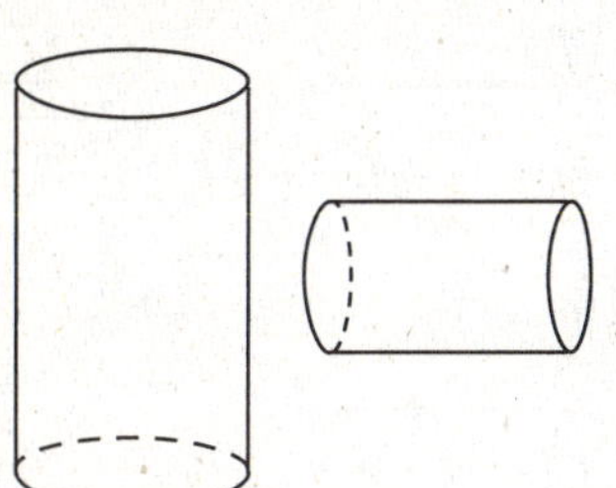

cylinder

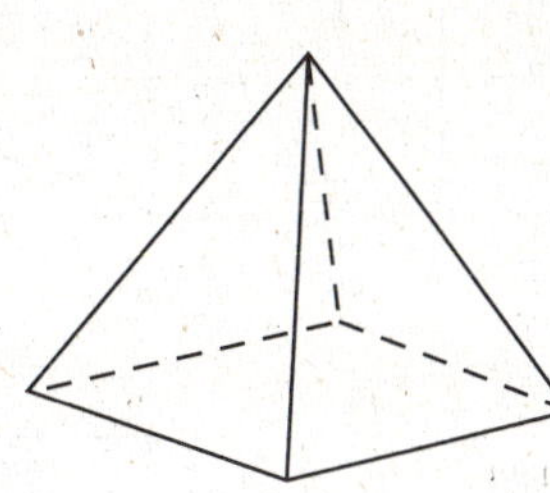

pyramid

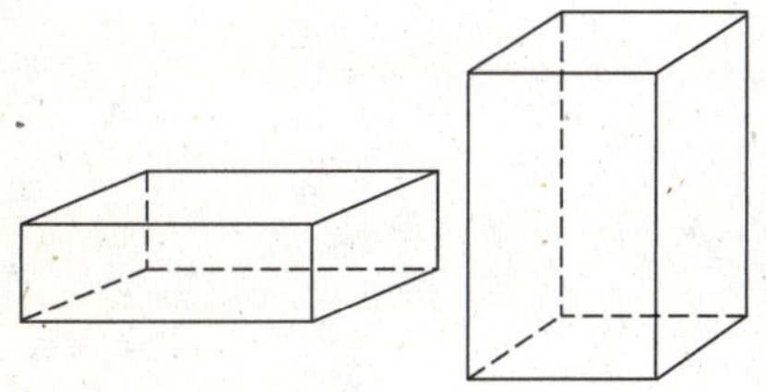

rectangular prism

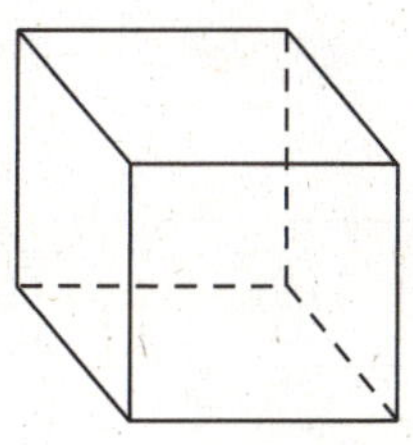

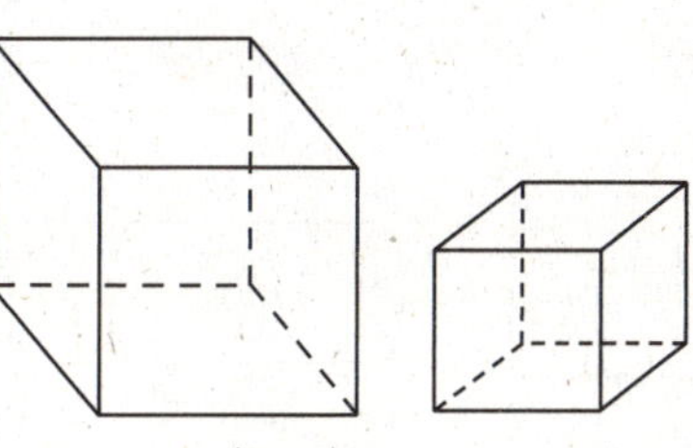

cube

Circle each pyramid.

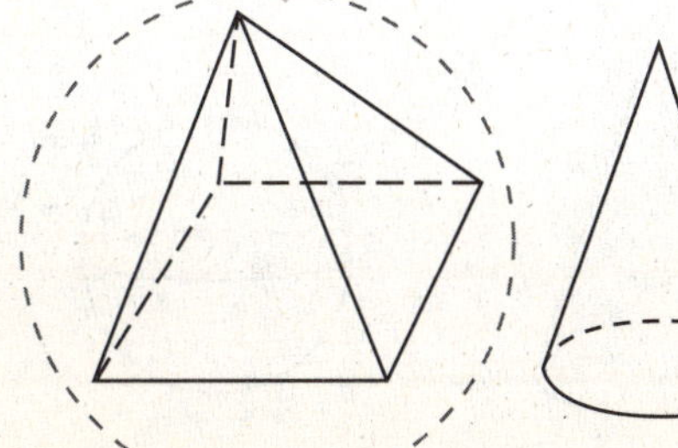

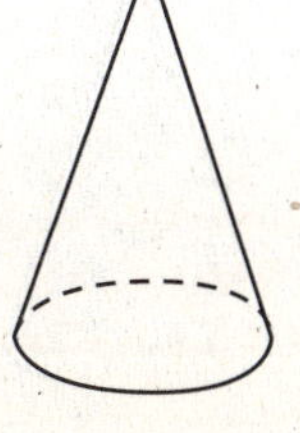

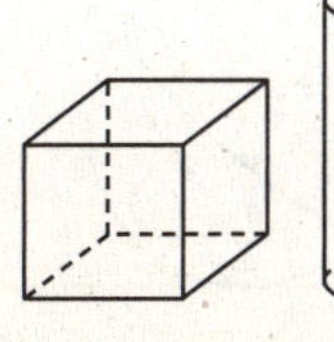

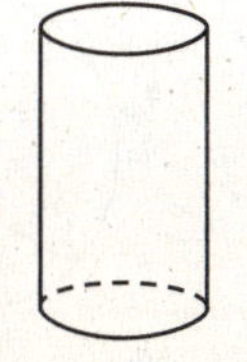

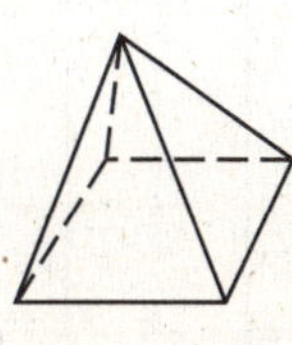

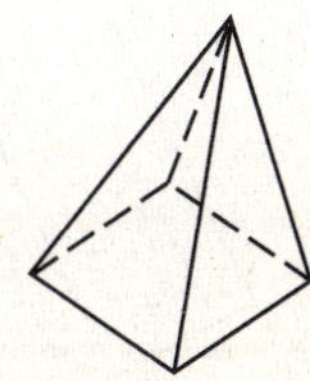

I. Circle each sphere.

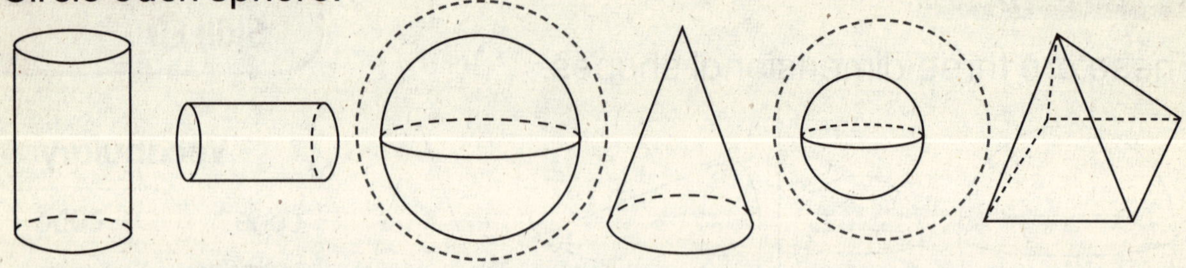

2. Circle each cylinder.

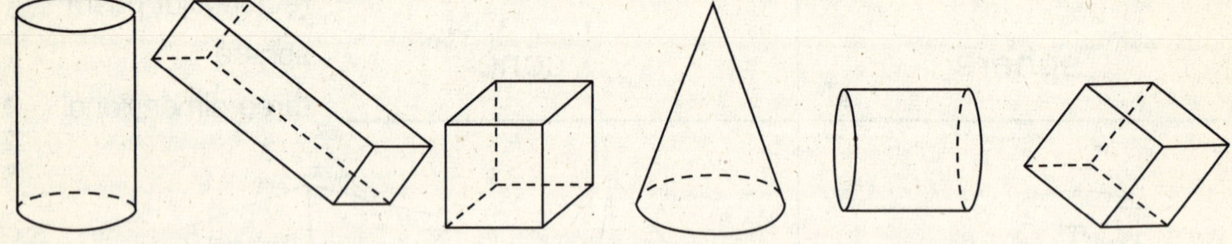

3. Circle each cube.

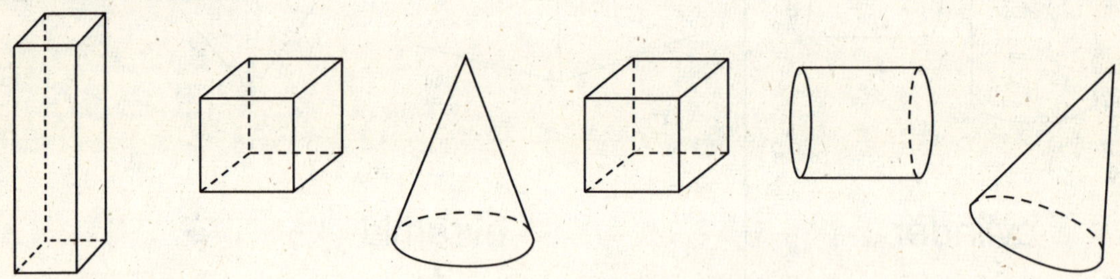

4. Circle each cone.

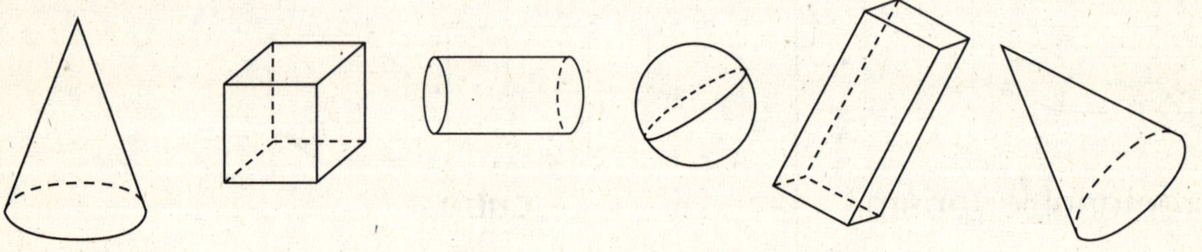

5. Circle each rectangular prism.

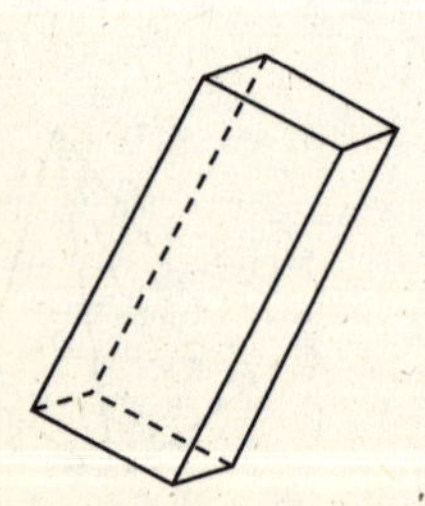
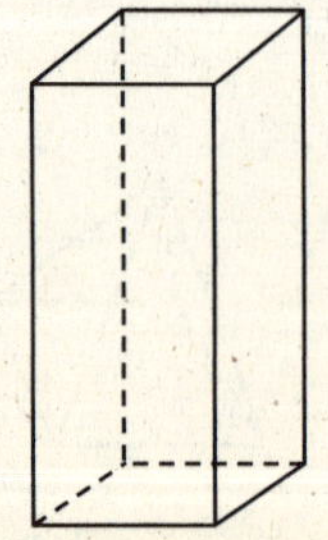
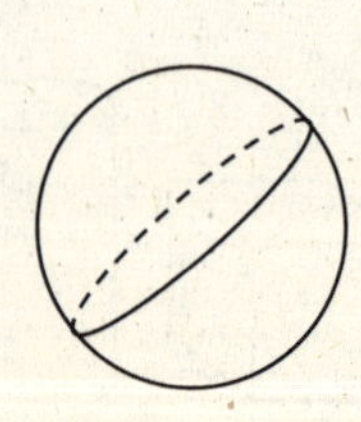
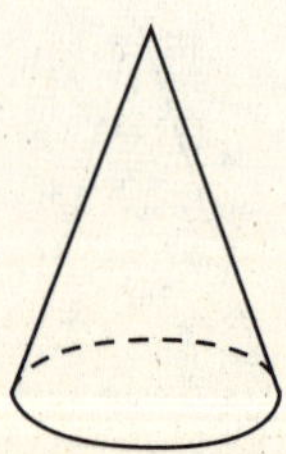
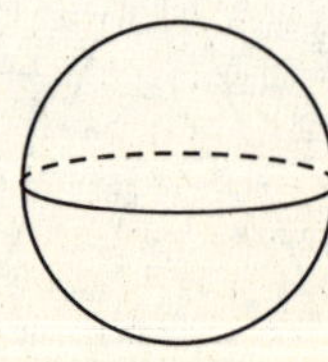

Name________________________________

Sides and Vertices
Skill 32

Learn the Math

Some shapes have sides and vertices.

Count the number of sides and vertices.

Vocabulary

side

vertex/vertices

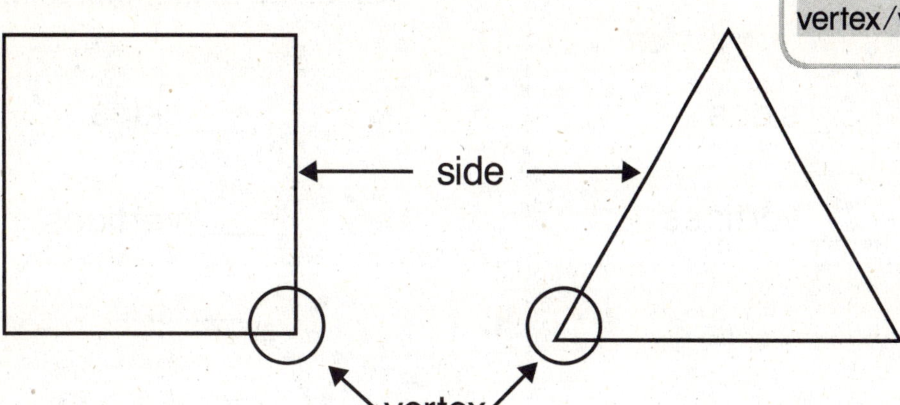

4 sides

4 vertices

3 sides

3 vertices

Trace each side. Circle each vertex.
Write how many sides and vertices.

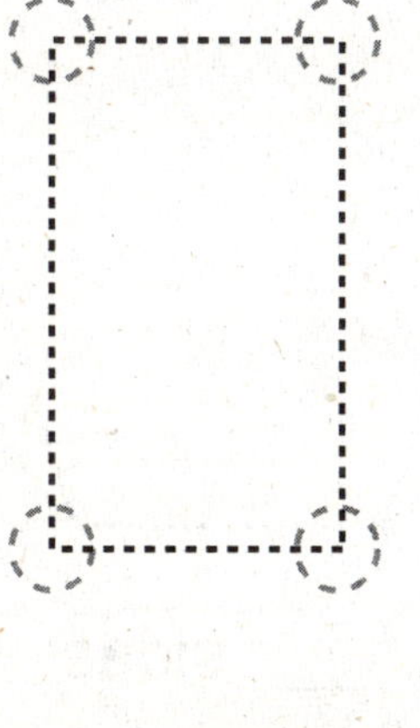

____ sides

____ vertices

____ sides

____ vertices

Trace each side. Circle each vertex. Write how many.

1.

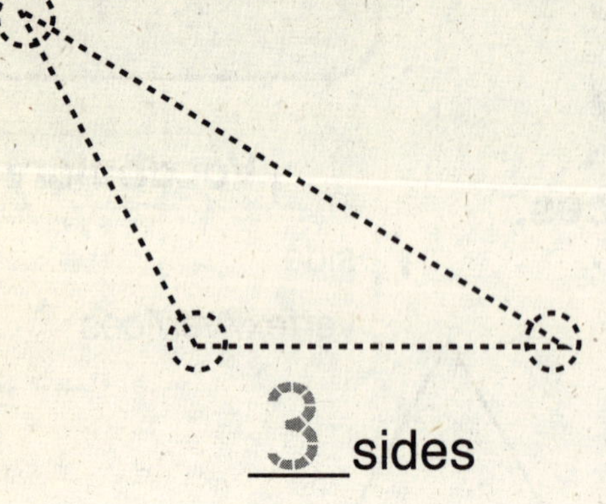

3 sides

3 vertices

2.

____ sides

____ vertices

3.

____ sides

____ vertices

4.

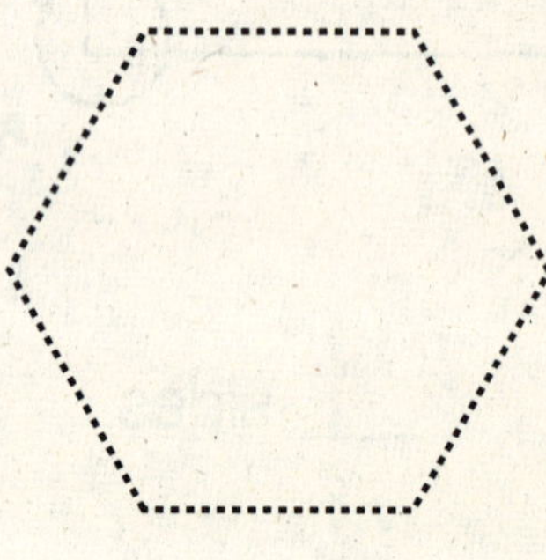

____ sides

____ vertices

5.

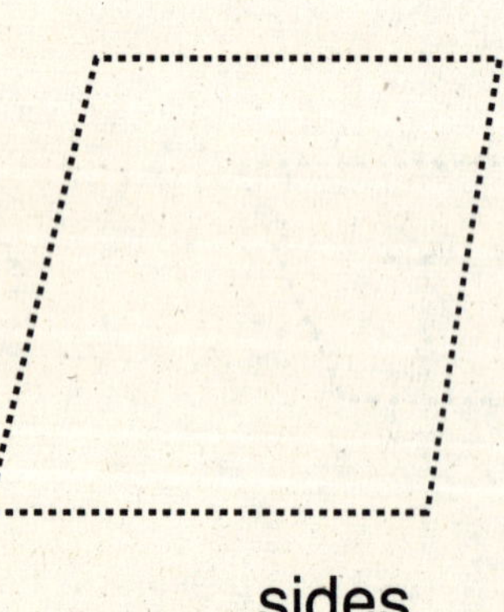

____ sides

____ vertices

6.

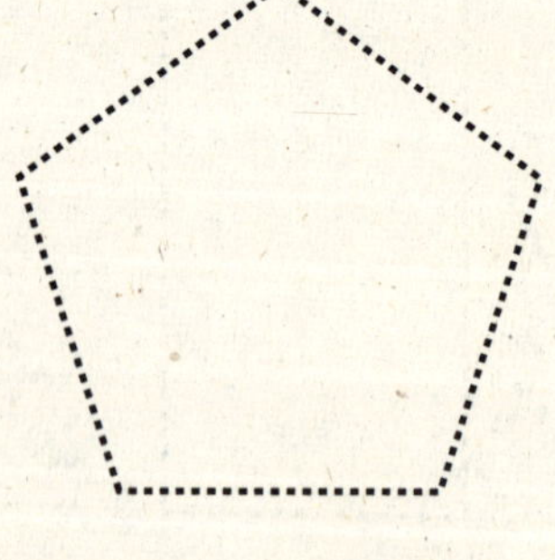

____ sides

____ vertices

Name____

Sort Two-Dimensional Shapes

Skill 33

Learn the Math

You can sort shapes by the number of sides and vertices they have.

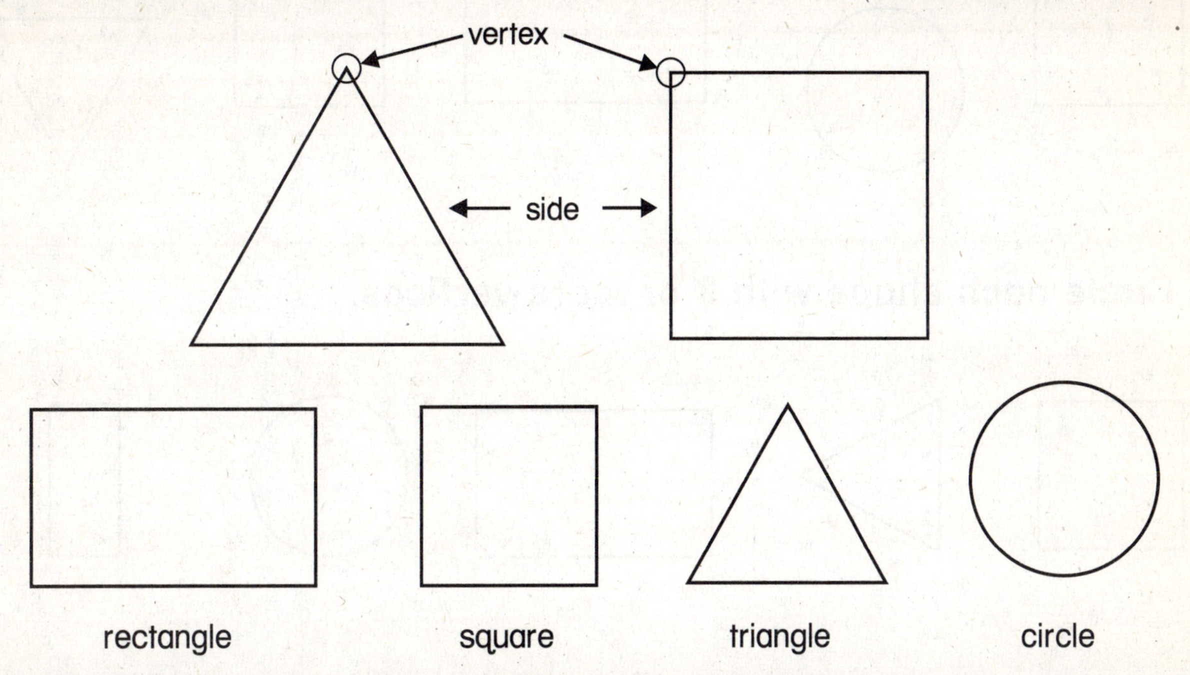

Circle each shape with 3 or more vertices.

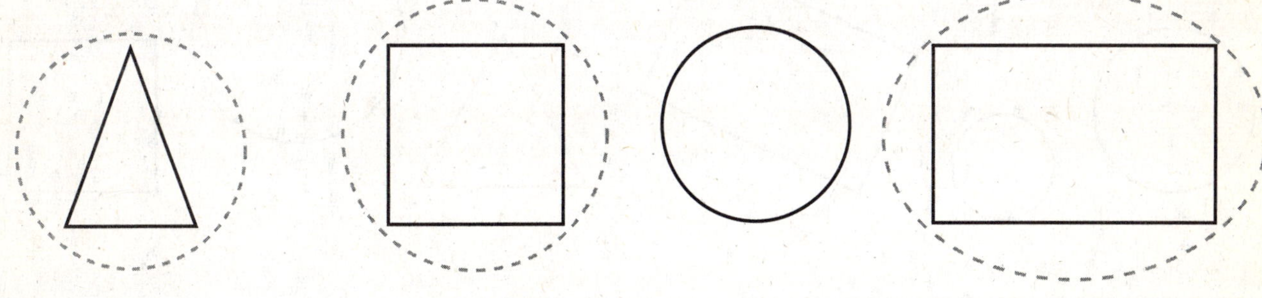

Mark an X on each triangle.

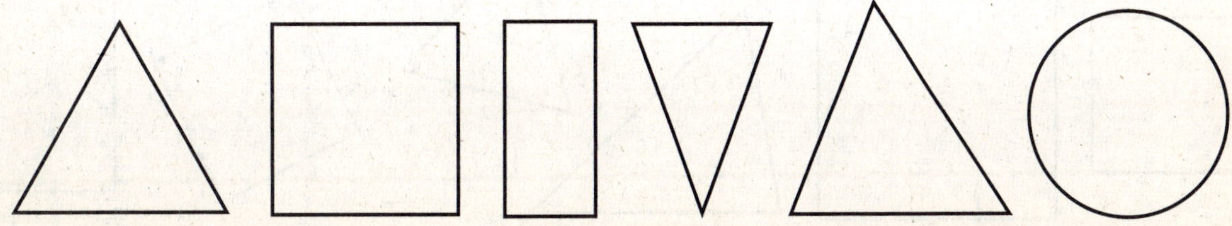

1. Circle each shape with 4 sides.

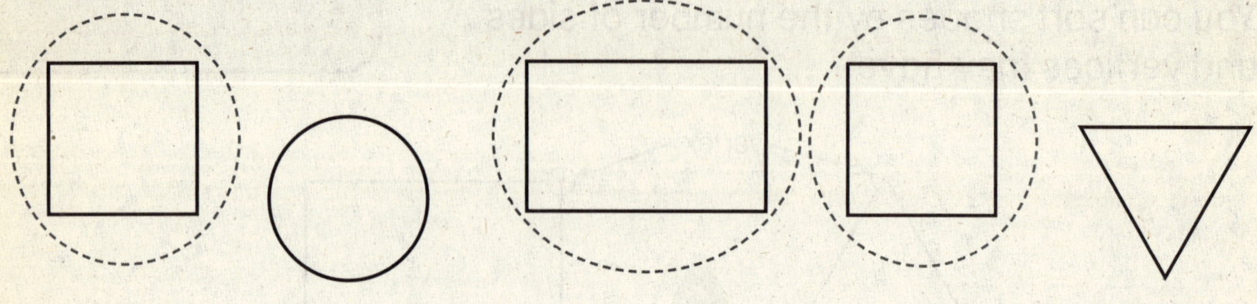

2. Circle each shape with 3 or more vertices.

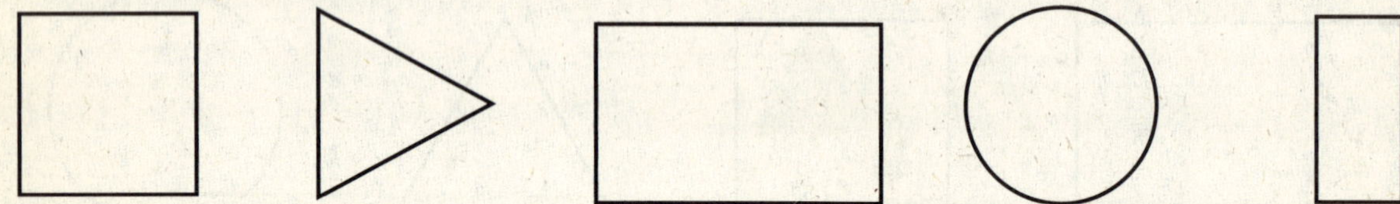

3. Mark an X on each circle.

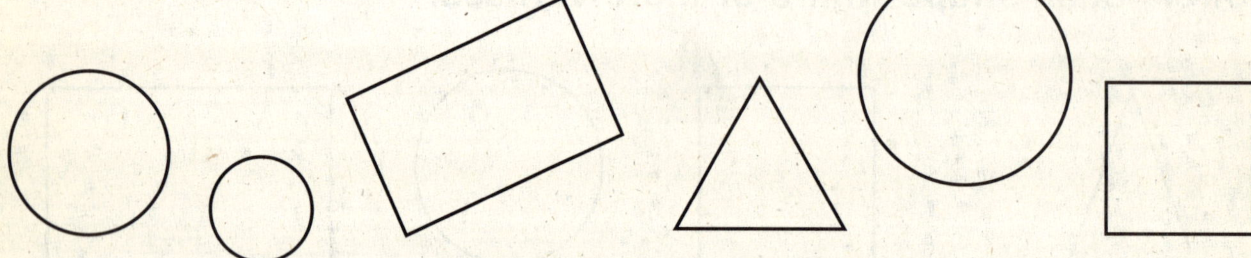

4. Mark an X on each square.

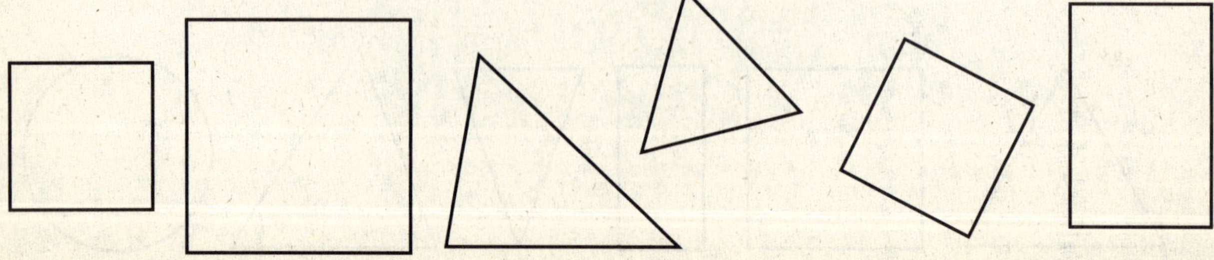

Name________________________________

Identify and Copy Patterns

Skill 34

Learn the Math

You can copy a pattern.

Vocabulary

pattern

Step 1 Look for the pattern.

Step 2 Draw shapes to match.

Step 3 Color the shapes to match.

Color the shapes to copy the pattern.

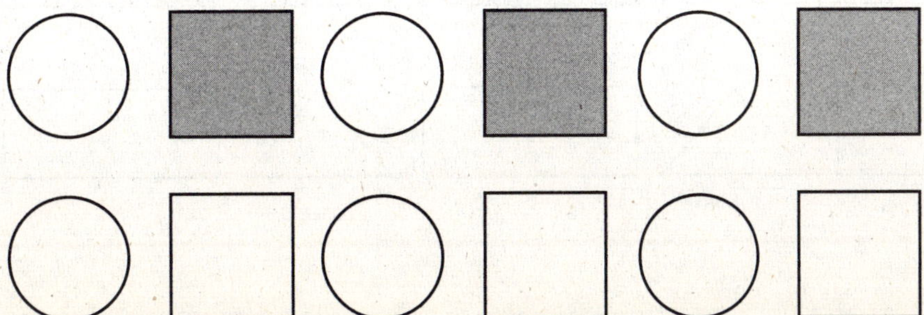

Color the shapes to copy the pattern.

1.

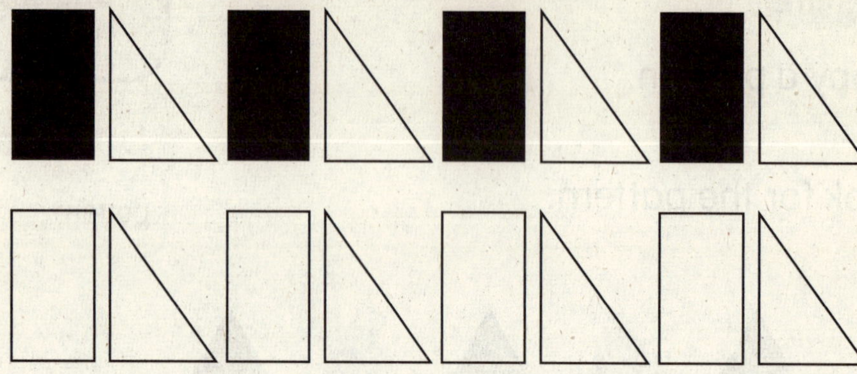

2.

3.

4.

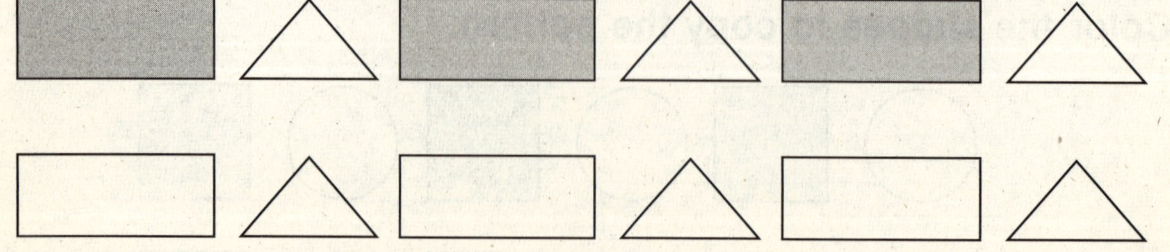

Name______________________________

Extend Patterns

Skill 35

Learn the Math

You can use a pattern unit to find what comes next in the pattern.

Vocabulary

pattern unit

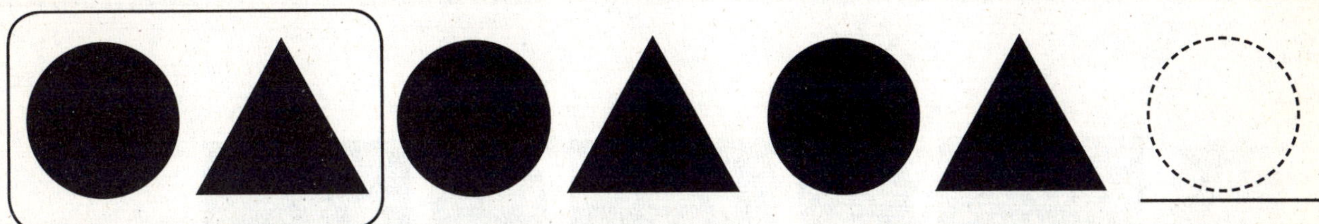

The pattern unit is circle, triangle.
Circle comes next.

Circle the pattern unit. Draw what comes next.

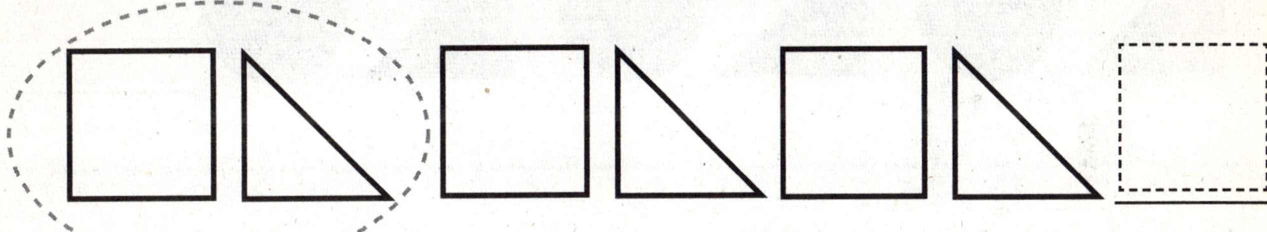

The pattern unit is square, triangle.
Square comes next.

The pattern unit is triangle, circle, square.
Triangle comes next.

Circle the pattern unit. Draw what comes next.

1.

2.

3.

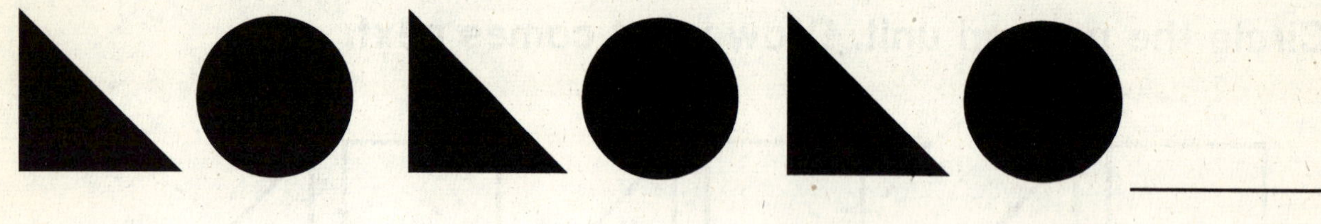

4.

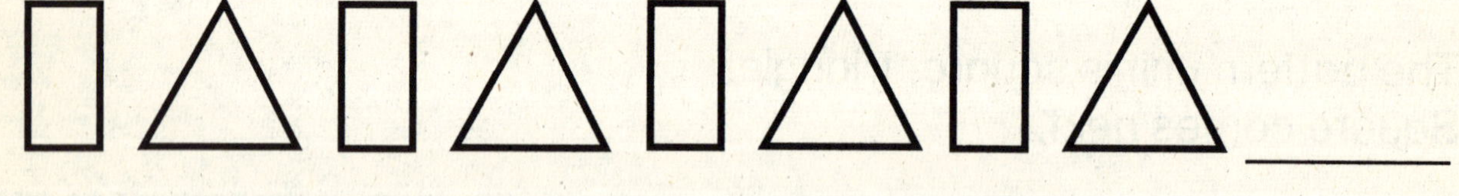

5.

Name________________________________

Compare Lengths

Skill 36

Learn the Math

You can compare lengths.

These objects are in order from shortest to longest.

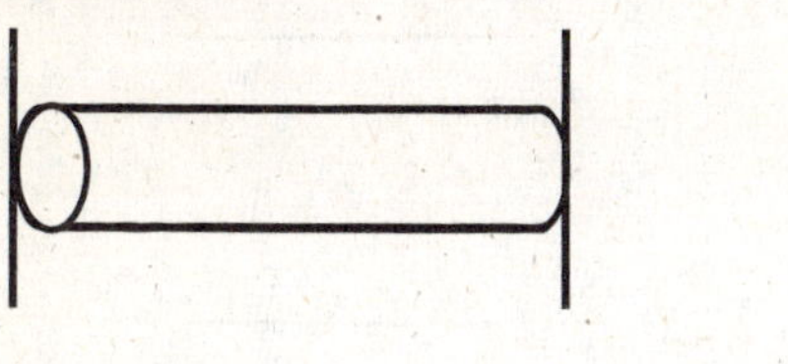

Vocabulary

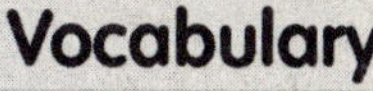

longest

shortest

shortest

longest

Circle the objects that are in order from shortest to longest.

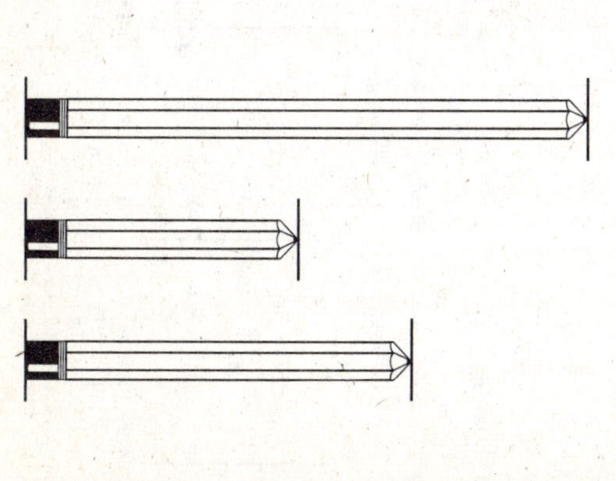

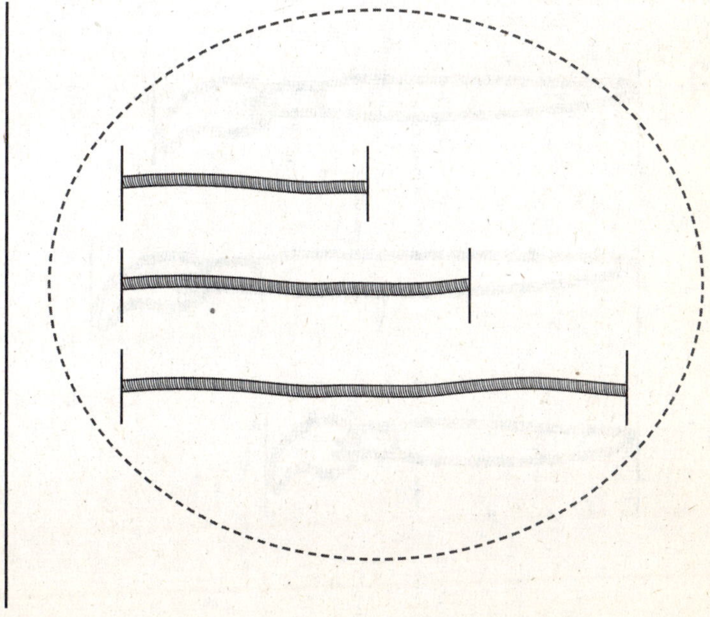

Order the objects from shortest to longest. Write 1, 2, and 3.

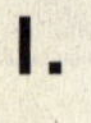

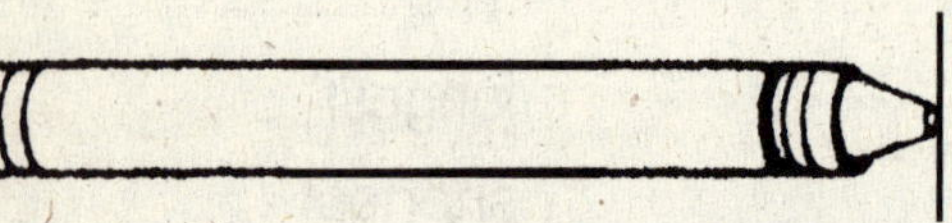

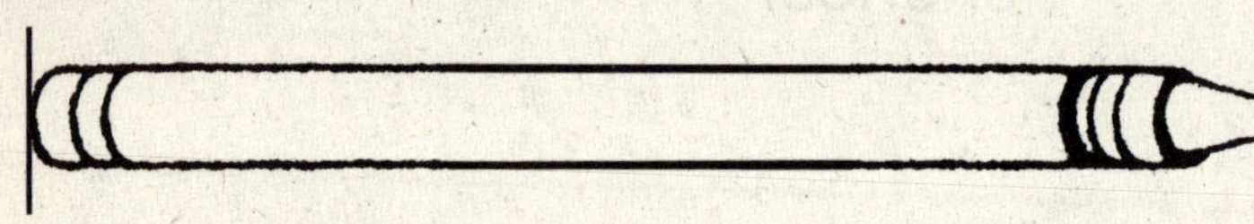

2.

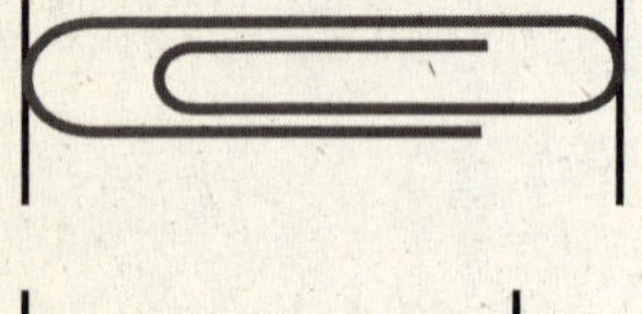

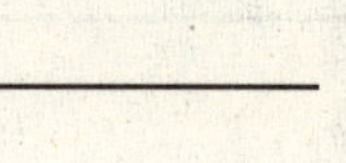

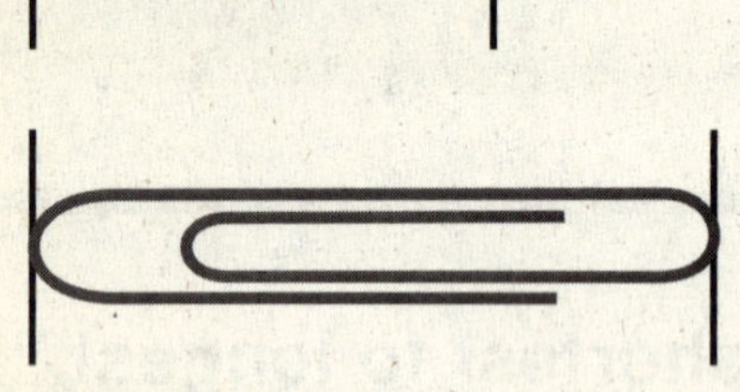

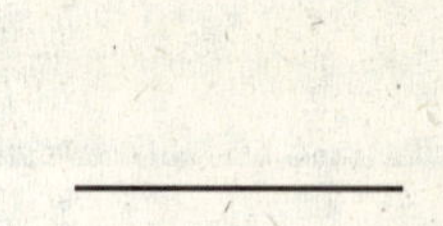

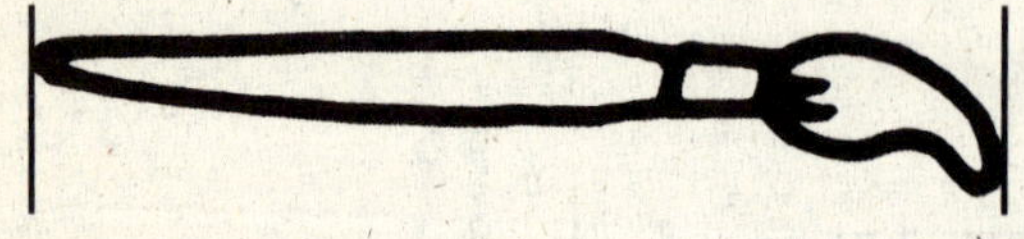

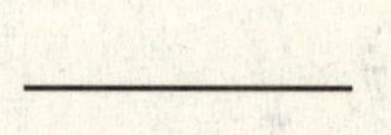

3.

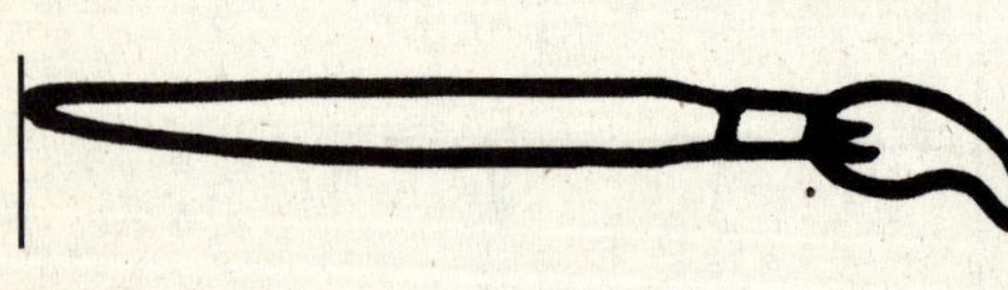

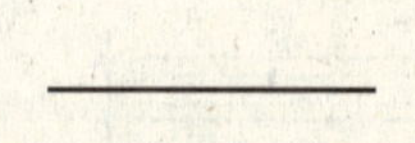

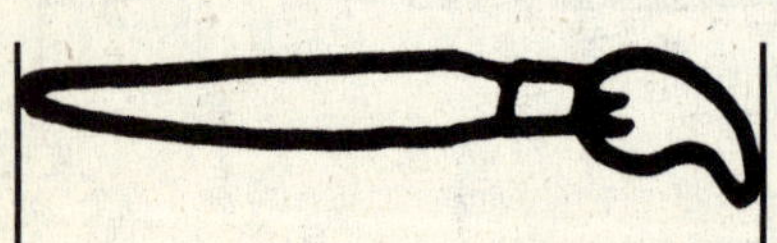

Name________________________________

Temperature

Skill 37

Learn the Math

Temperature is the measure of how hot or cold something is. A thermometer measures temperature.

Vocabulary

temperature

thermometer

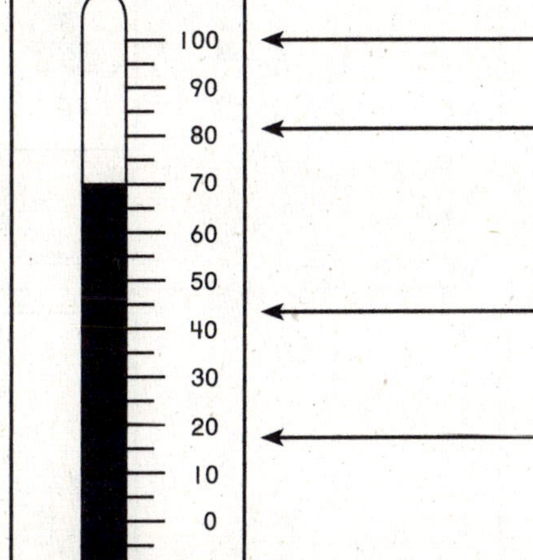

A very hot day is 100°F.

A warm day is 80°F.

A cool day is 45°F.

A very cold day is 18°F.

The temperature is 70 °F.

Read each thermometer.
Write the temperature.

1.

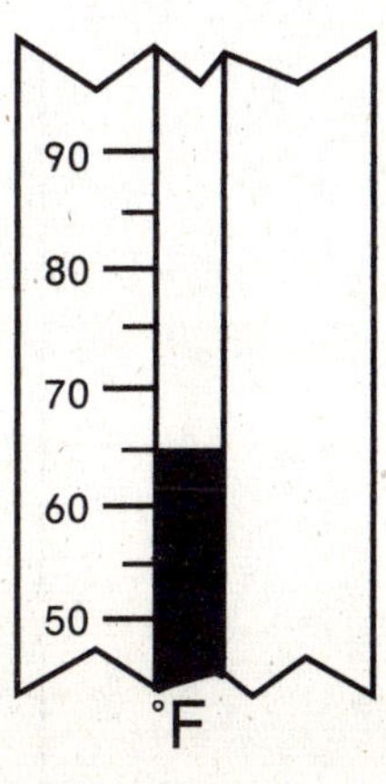

_____ °F

2.

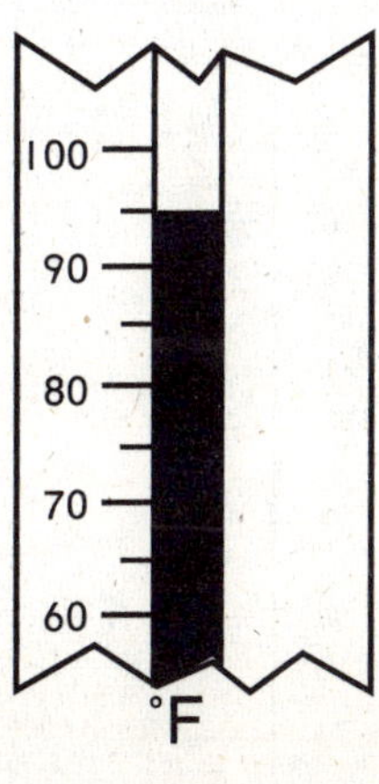

_____ °F

Read each thermometer.
Write the temperature.

1. 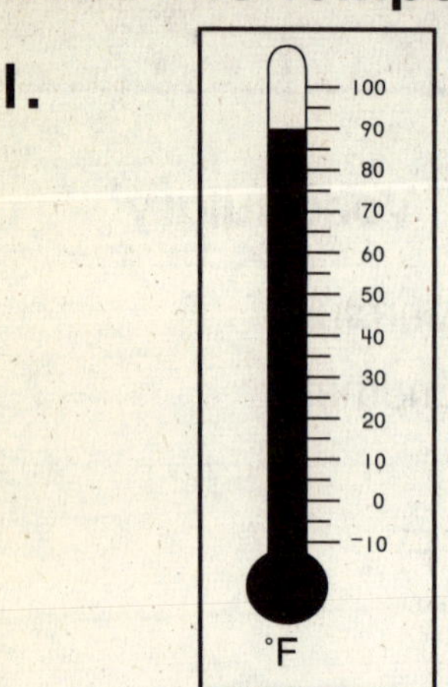

90 °F

2. 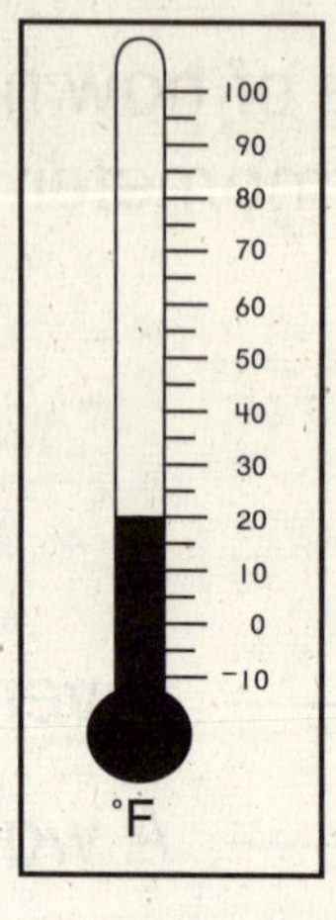

____ °F

3. 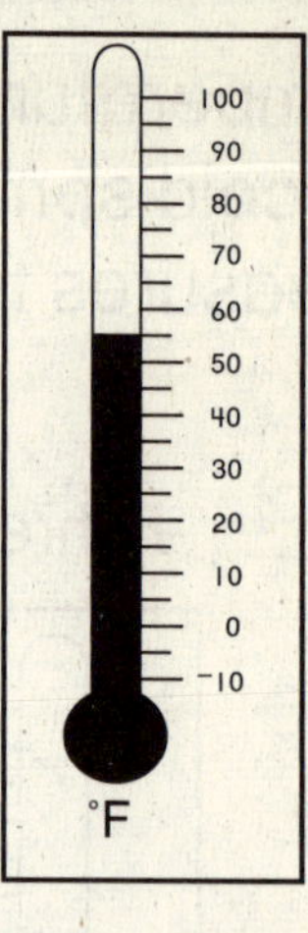

____ °F

4. 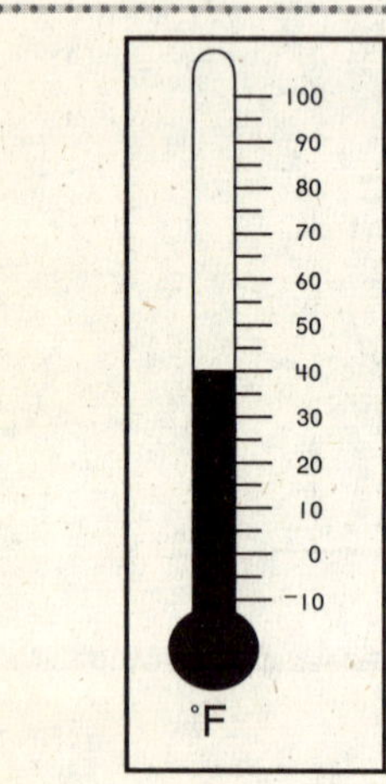

____ °F

5. 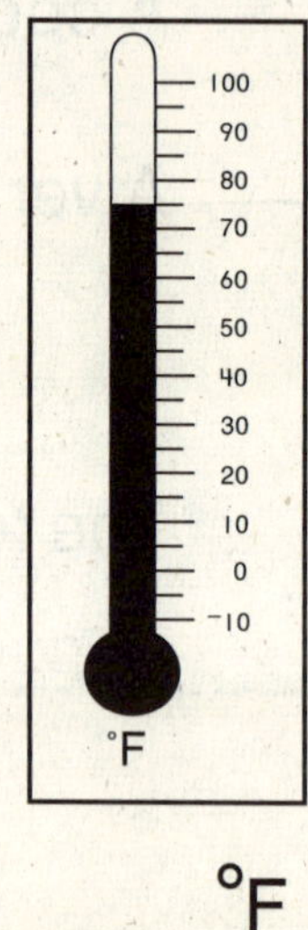

____ °F

6. 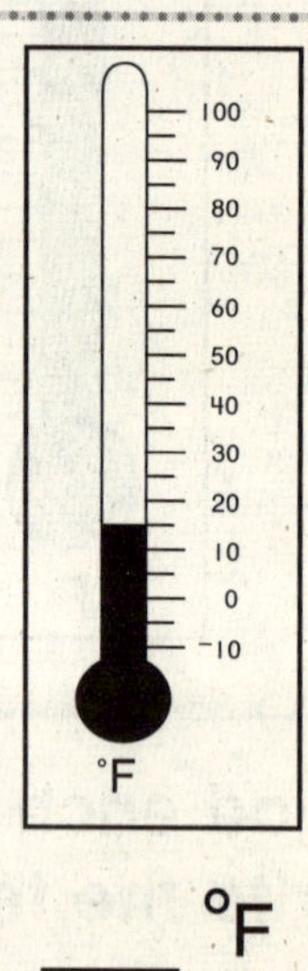

____ °F

7. 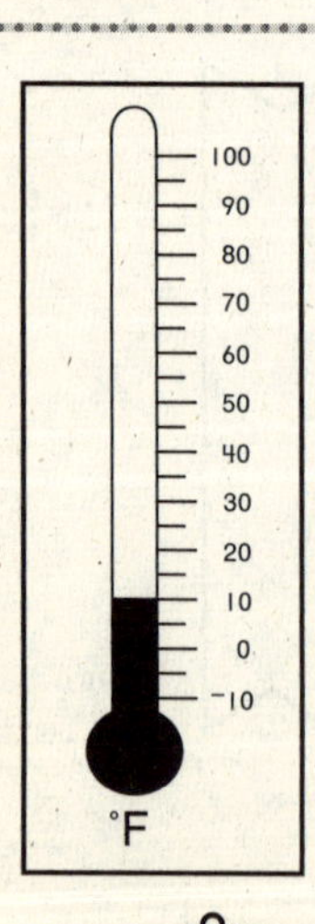

____ °F

8. 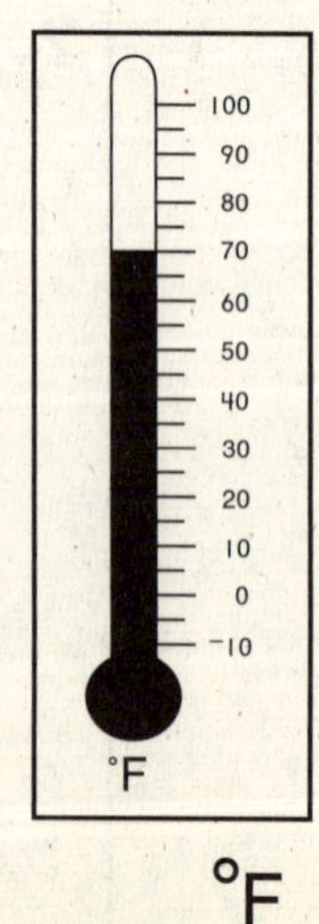

____ °F

9. 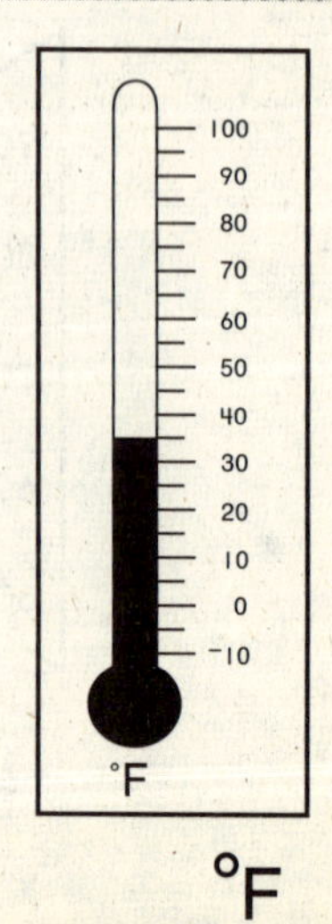

____ °F

Name_______________________________

Compare Weights

Skill 38

Learn the Math

You can hold objects to compare their weights.

Vocabulary
heavier
lighter

Circle the object that is heavier.

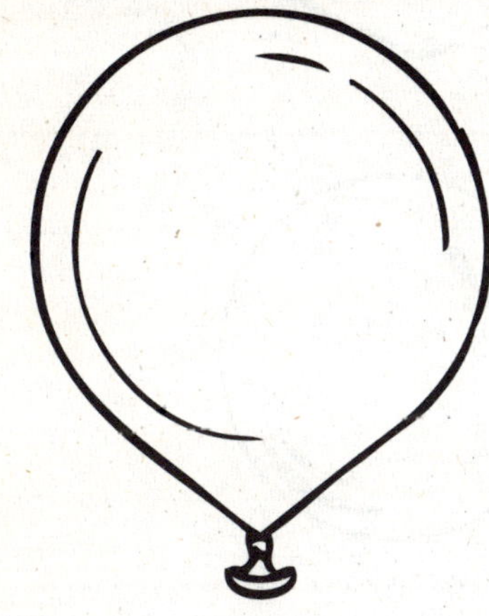

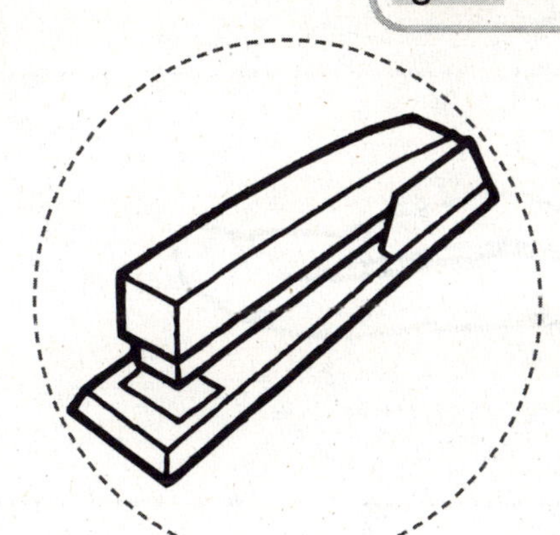

Hold one object in each hand.

Compare their weights.

The stapler is heavier.

Circle the object that is heavier.

Mark an X on the object that is lighter.

Circle the object that is heavier.
Mark an X on the object that is lighter.

1.

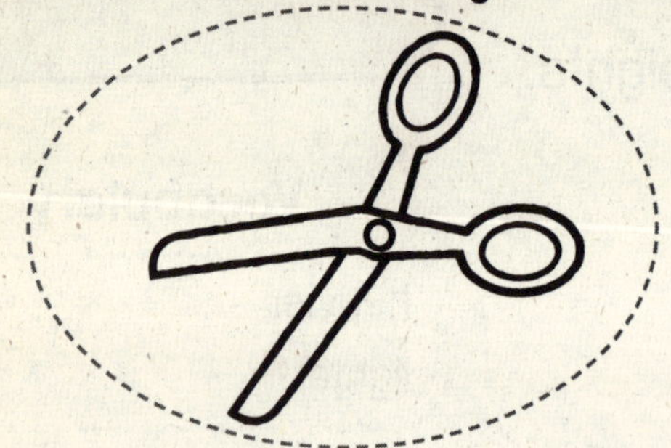

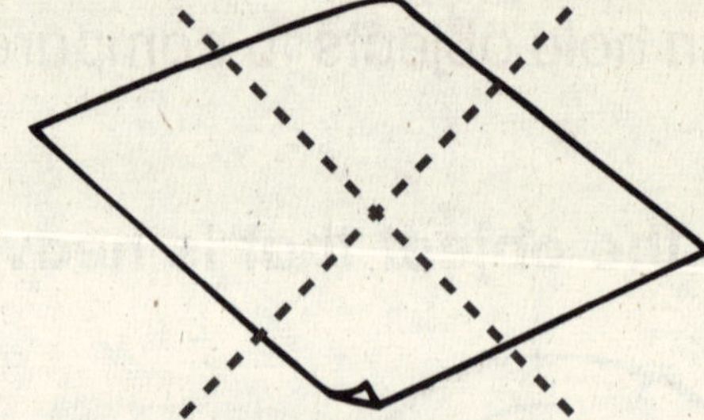

2.

3.

4.

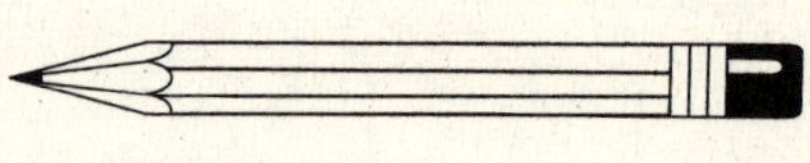

5.

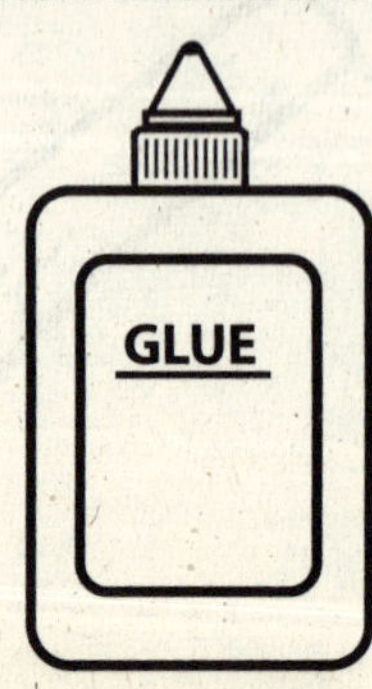

Name________________________________

Compare Capacities

Skill 39

Vocabulary

capacity
less
more

Learn the Math

You can compare containers by how much they hold.

Circle the container that holds more.

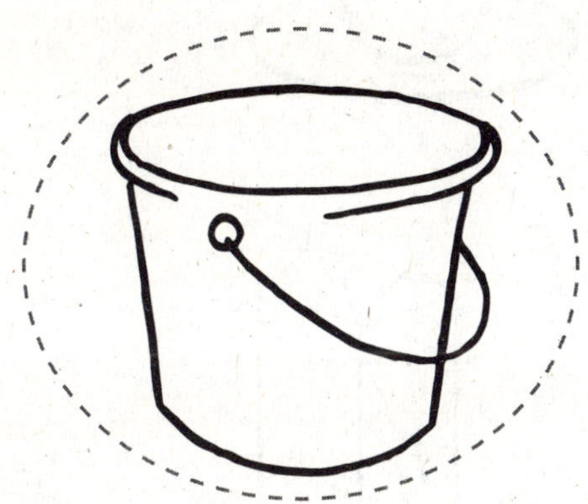

The  holds more than the .

Circle the container that holds more.

Circle the container that holds less.

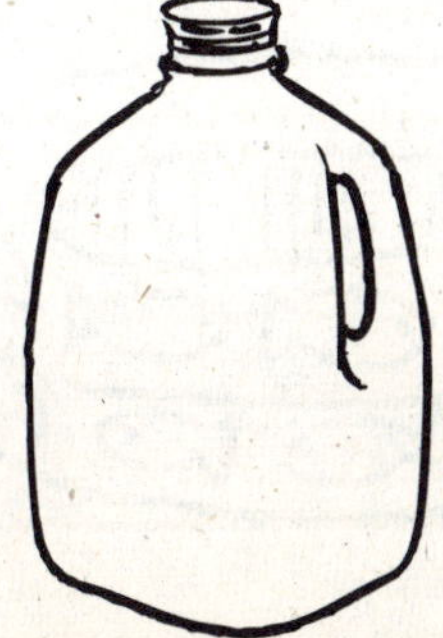

Circle the container that holds more.

1.

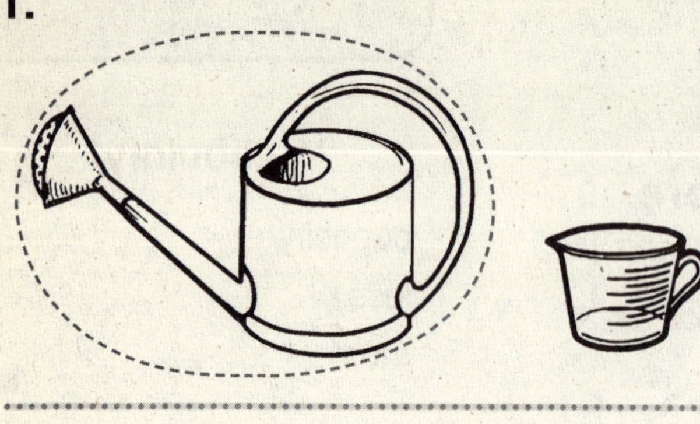

2.

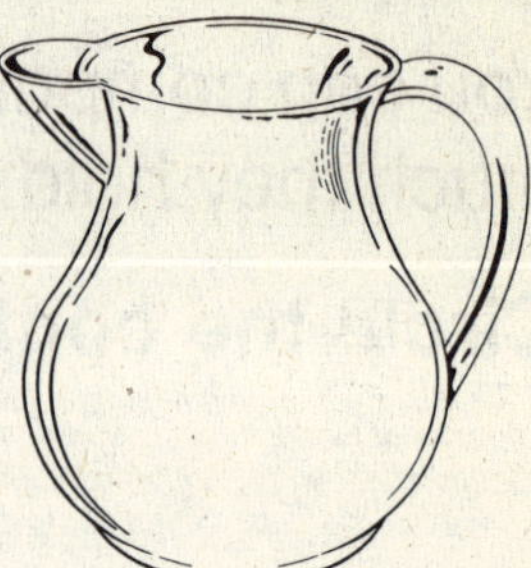

3.

4.

Circle the container that holds less.

5.

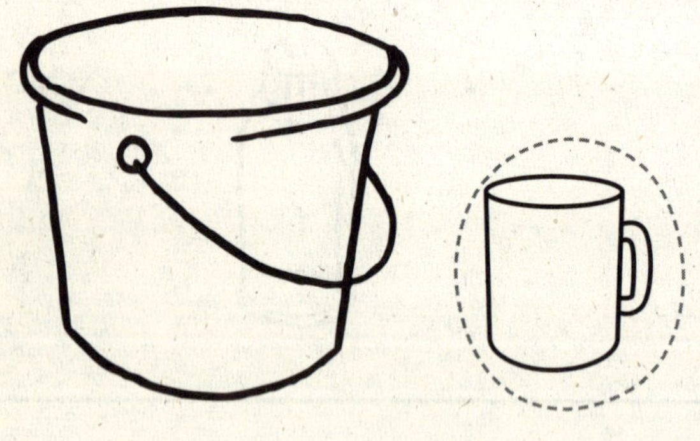

6.

7.

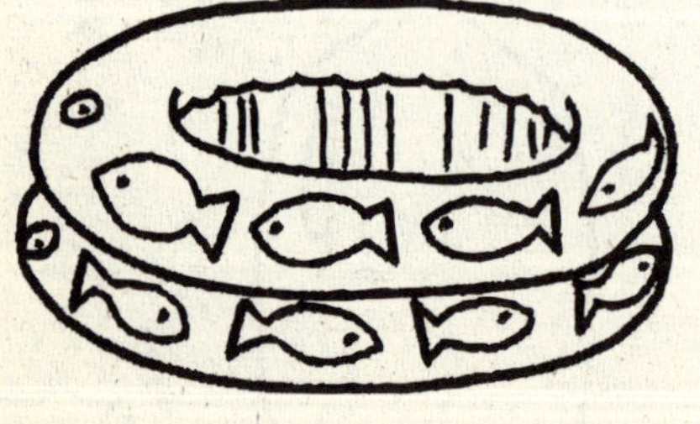

8.

Name____________________

Equal Parts
Skill 40

Learn the Math

Equal parts are the same size.

Vocabulary

equal parts

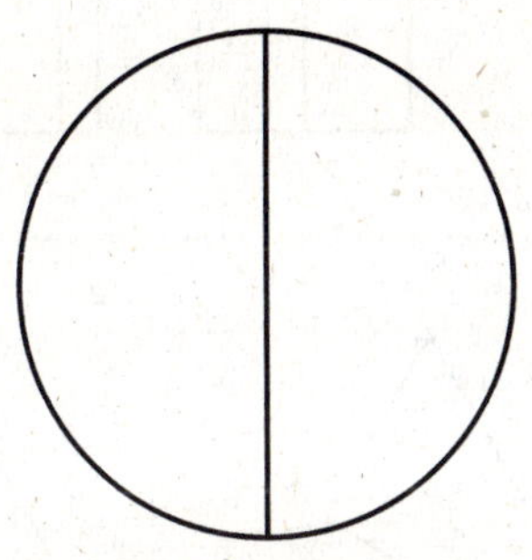
equal parts

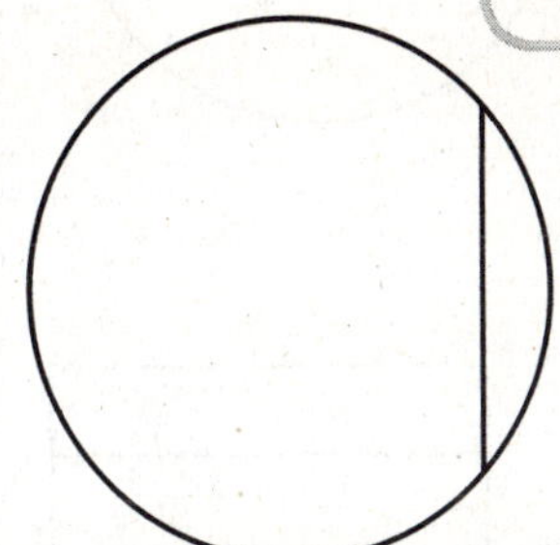
unequal parts

Circle the shape that has 2 equal parts.

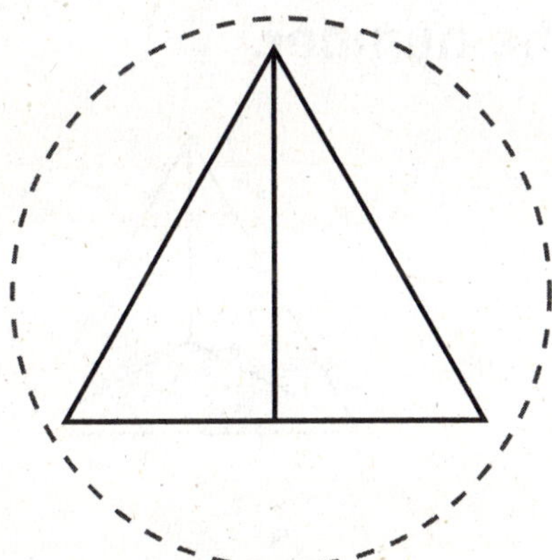

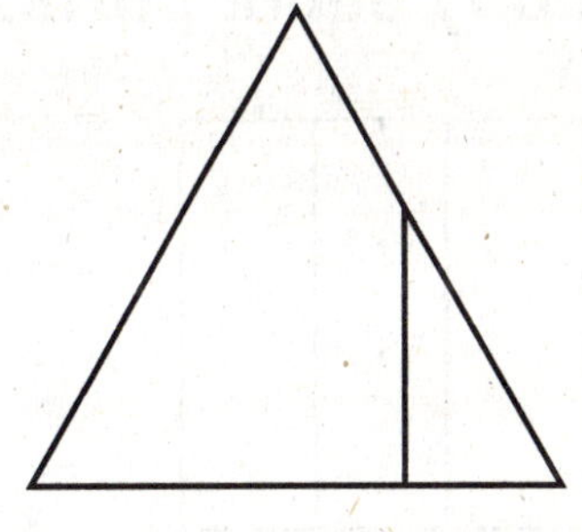

Count the equal parts.

2 equal parts

____ equal parts

____ equal parts

Circle the shape that has two equal parts.

1.

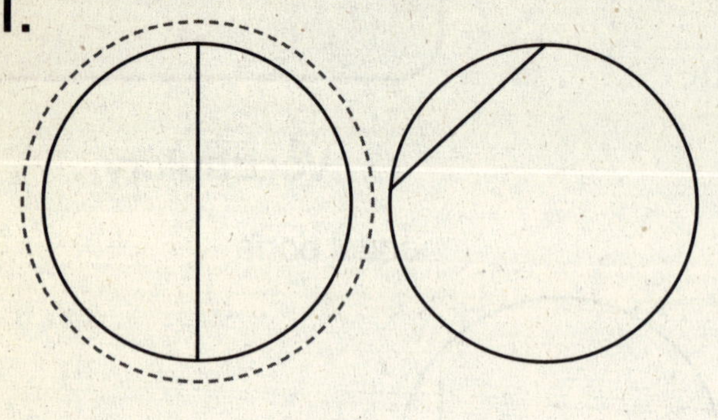

2.

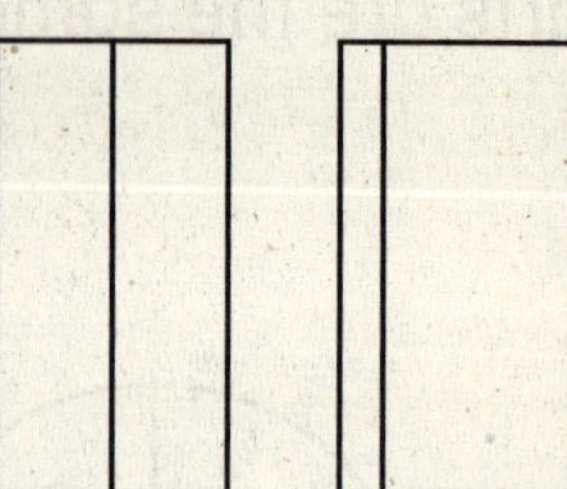

3.

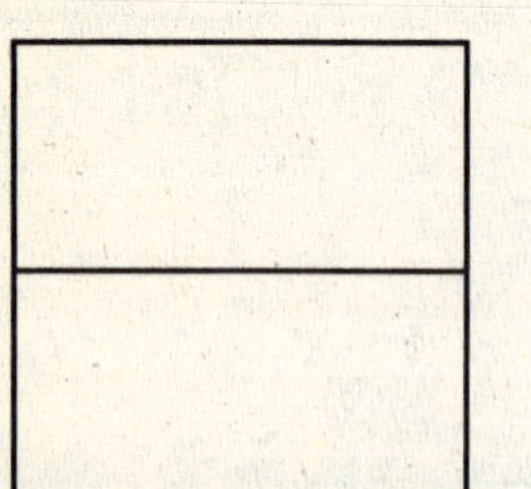

4.

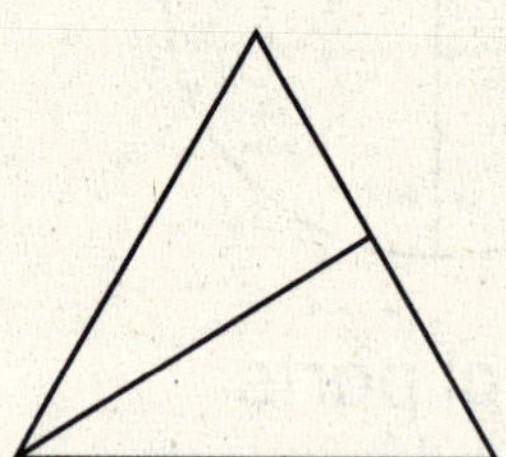

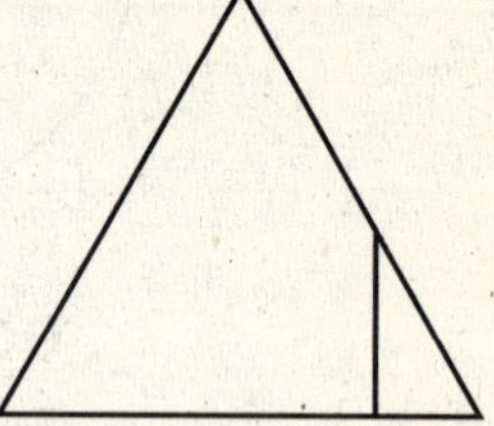

Count how many equal parts. Write the number.

5.

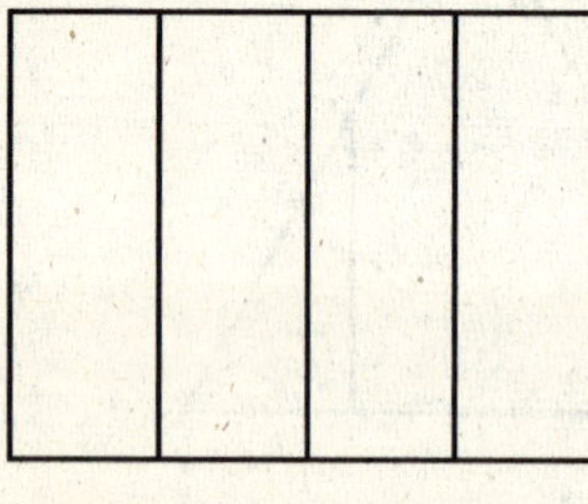

4 equal parts

6.

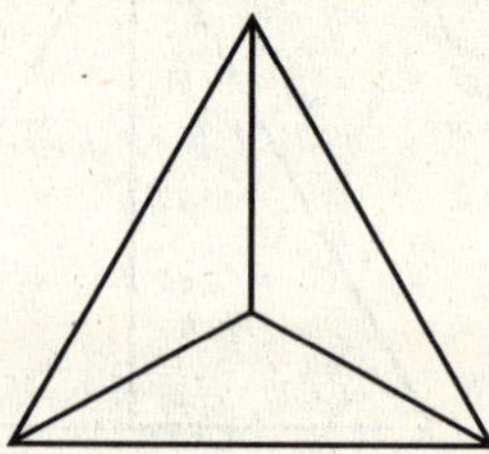

____ equal parts

7.

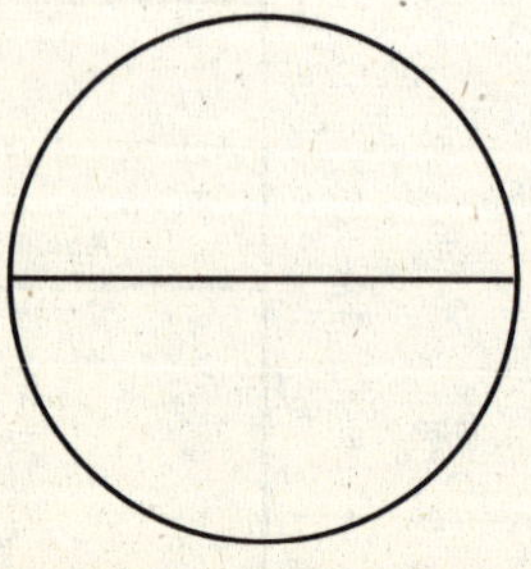

____ equal parts

8.

____ equal parts

Name________________________________

Explore Halves

Skill 41

Learn the Math

Each of 2 equal parts of a whole is one half.

Vocabulary

one half $\frac{1}{2}$

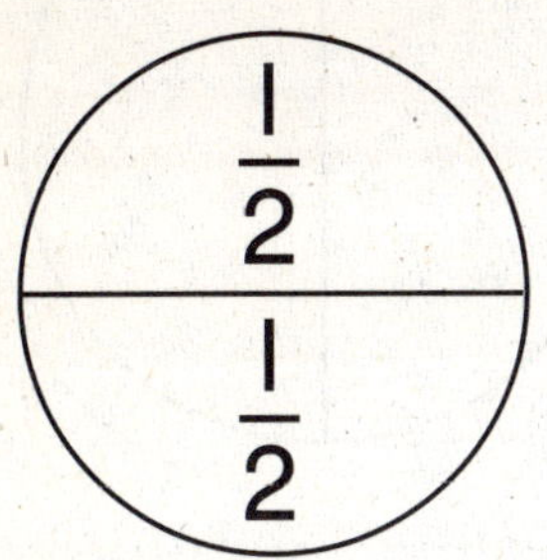

Two halves make one whole.

1 of 2 equal parts is $\frac{1}{2}$ or one half.

Circle the shape that shows halves.

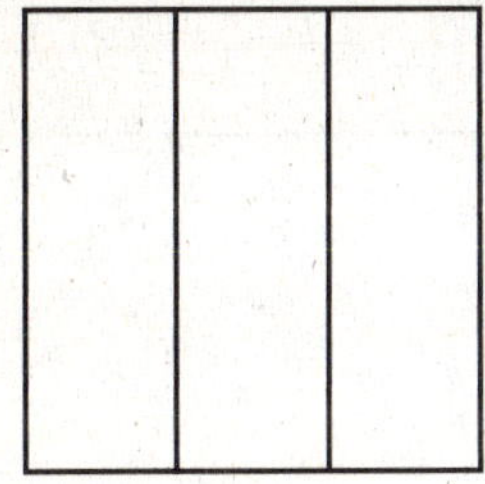

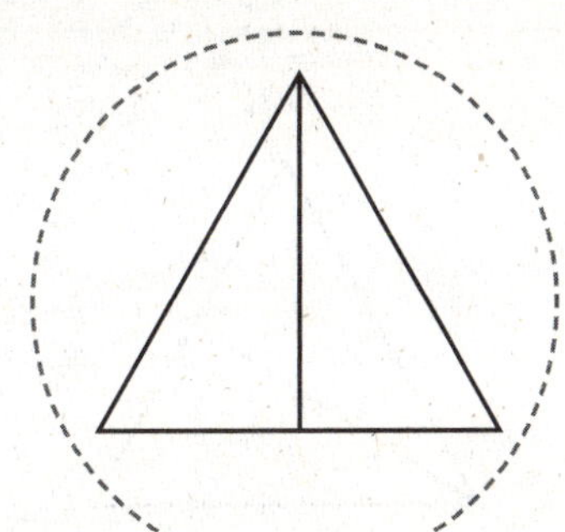

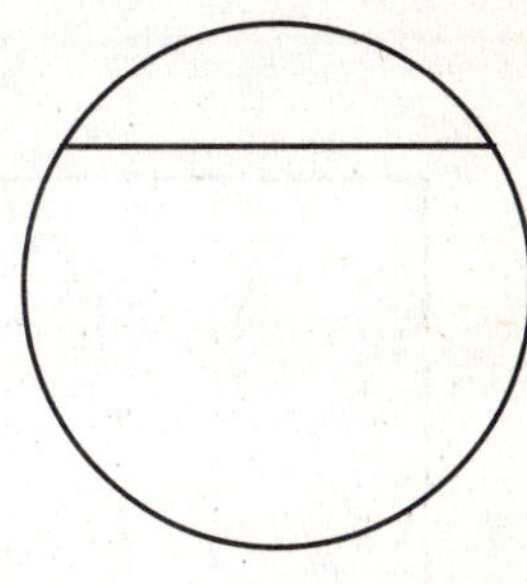

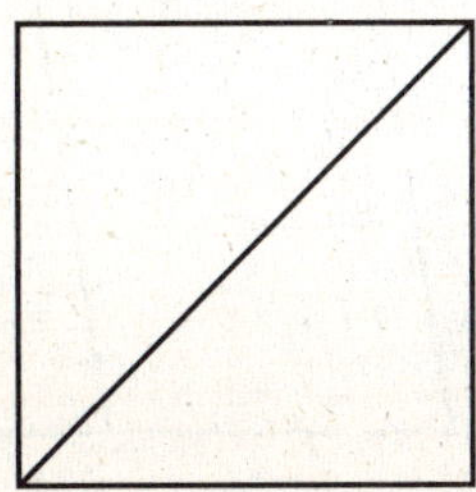

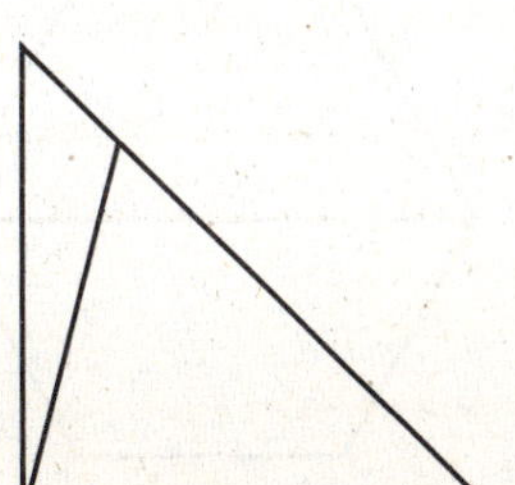

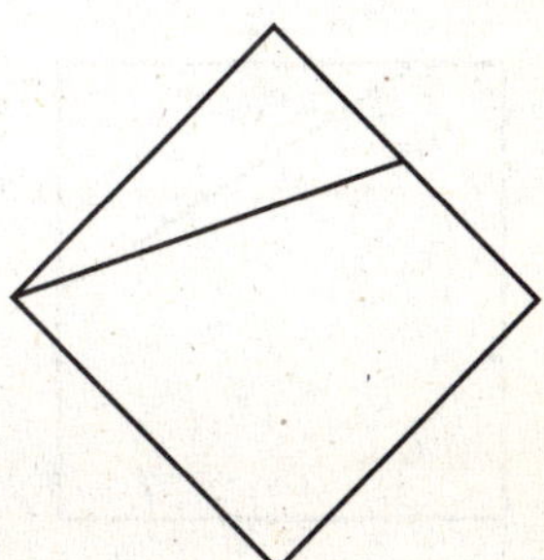

Circle the shape that shows halves.

1.

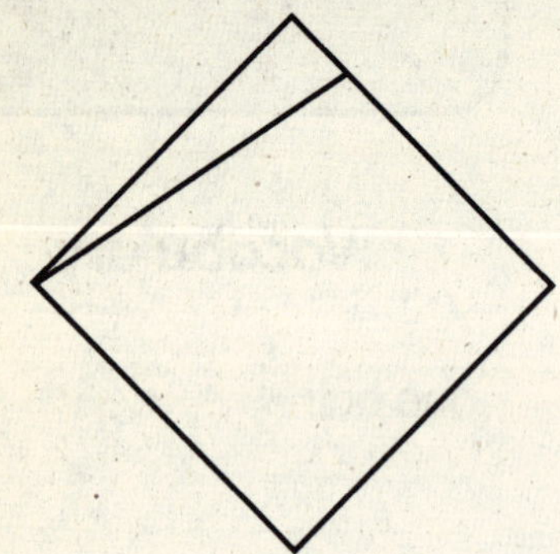

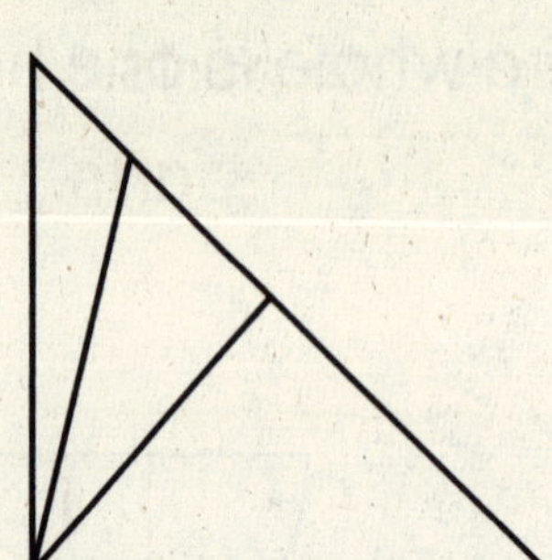

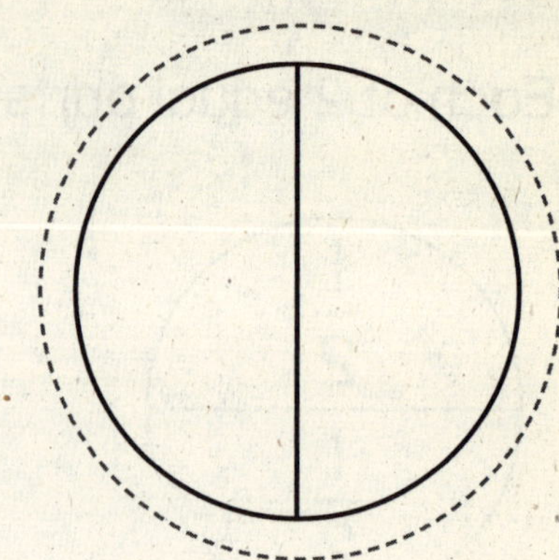

2.

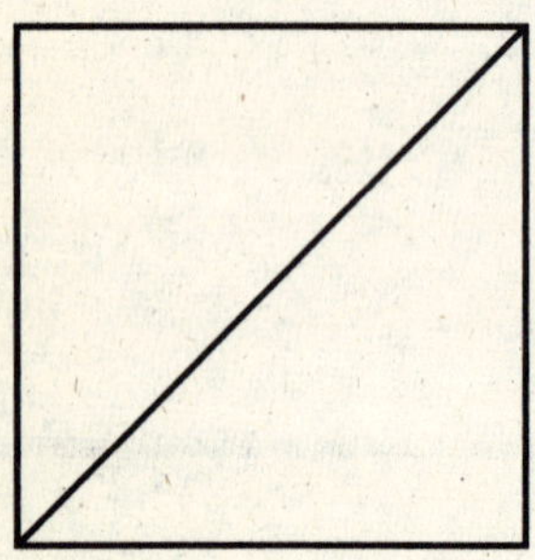

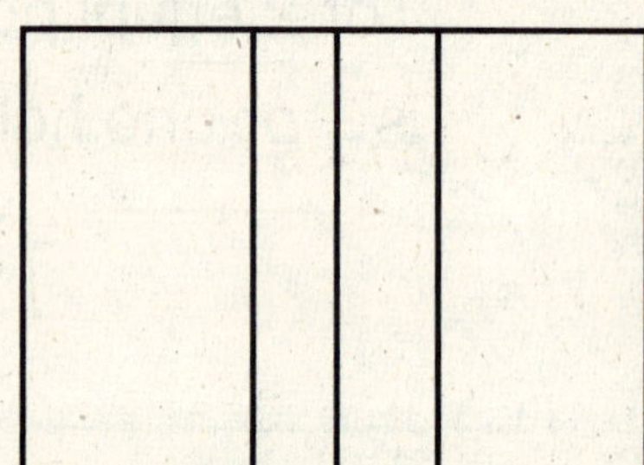

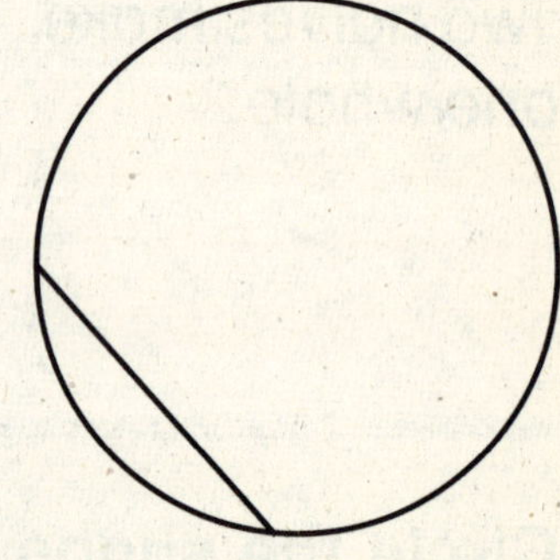

3.

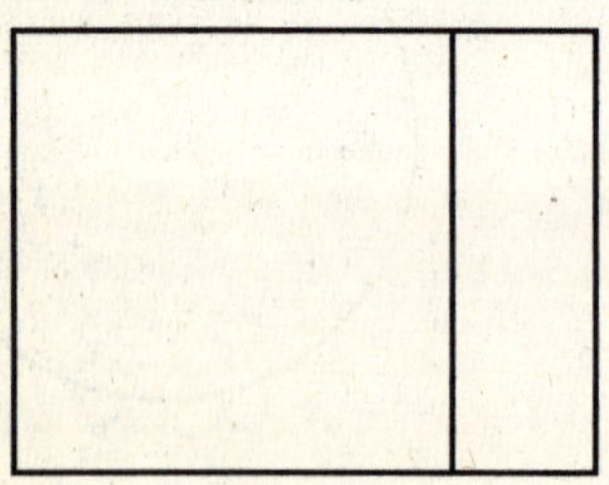

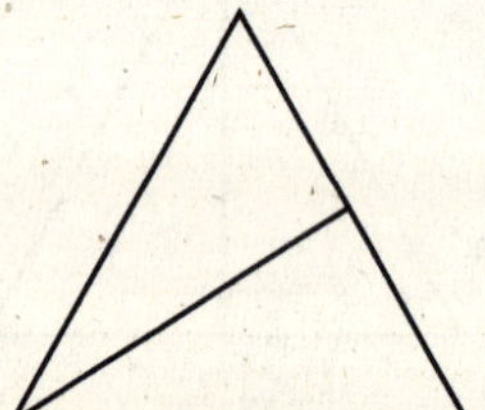

4.

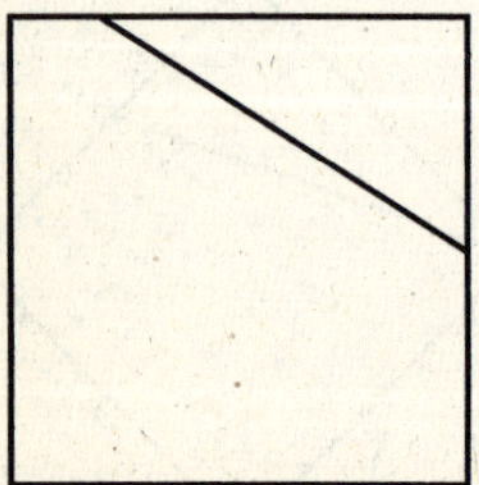

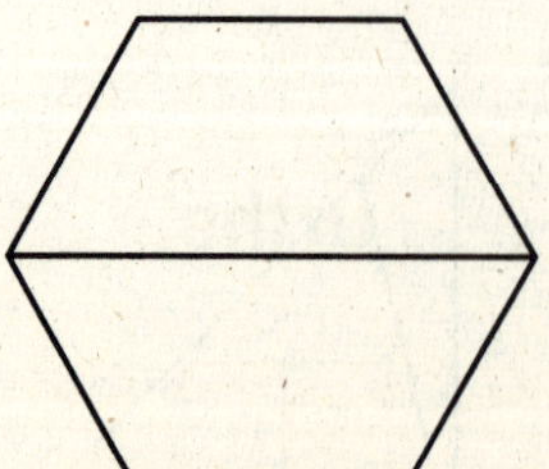

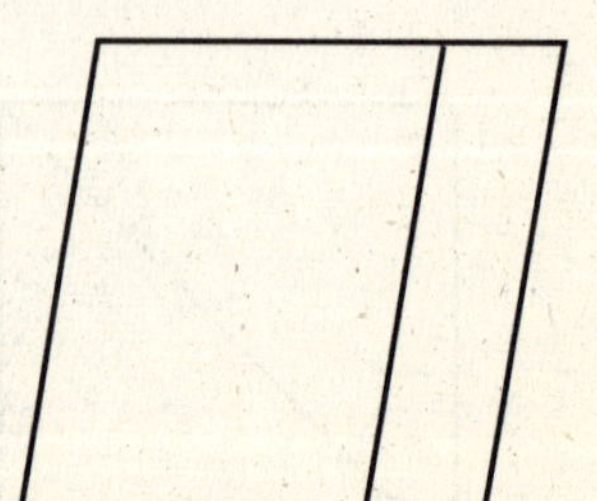

Name___________________________________

Explore Thirds and Fourths

Skill 42

Learn the Math

A shape with thirds has 3 equal parts.

A shape with fourths has 4 equal parts.

1 of 3 equal parts is $\frac{1}{3}$ or one third.

Vocabulary

one third $\frac{1}{3}$

one fourth $\frac{1}{4}$

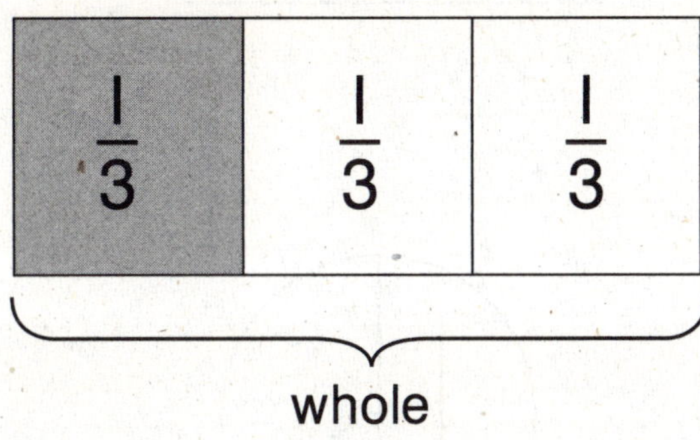

1 of 4 equal parts is $\frac{1}{4}$ or one fourth.

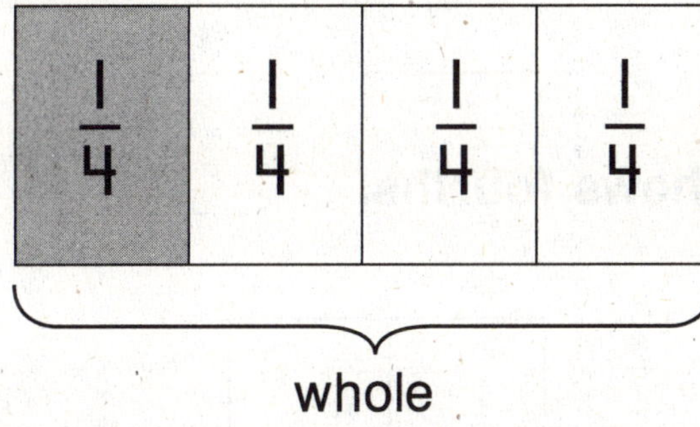

Circle the shape that shows thirds.

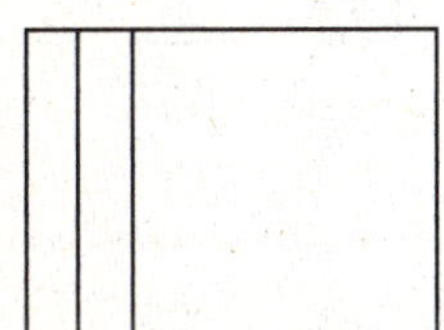
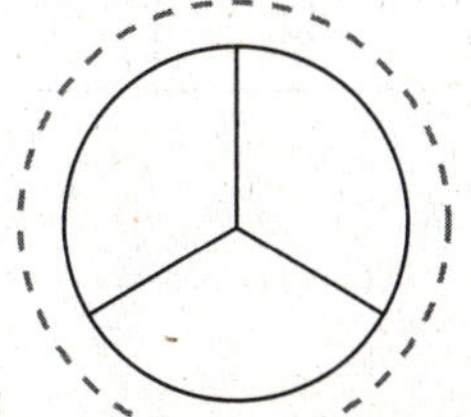
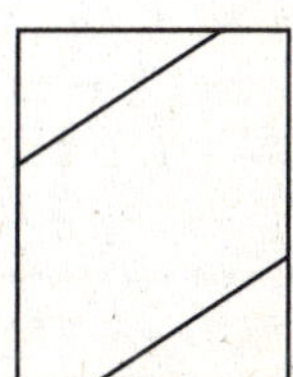

Circle the shape that shows fourths.

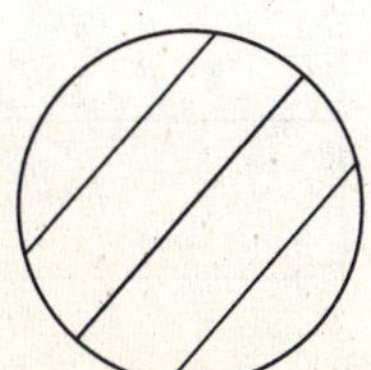
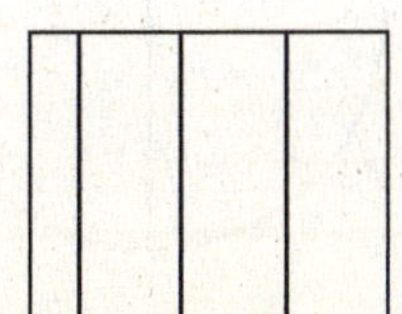
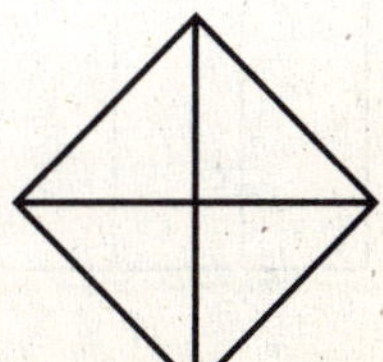

Circle the shape that shows thirds.

1.

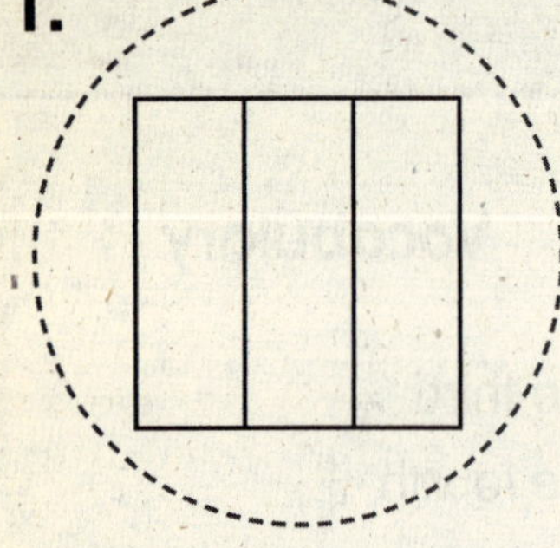

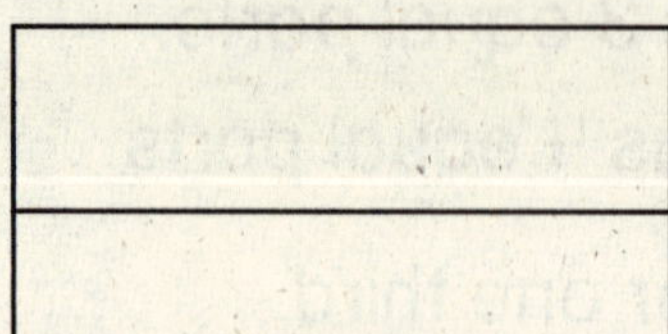

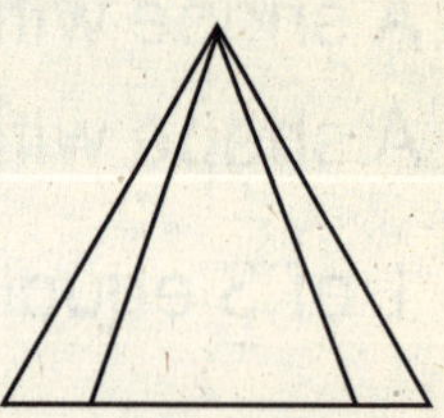

2.

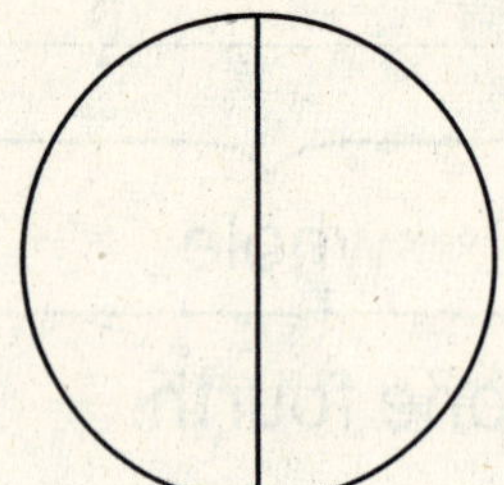

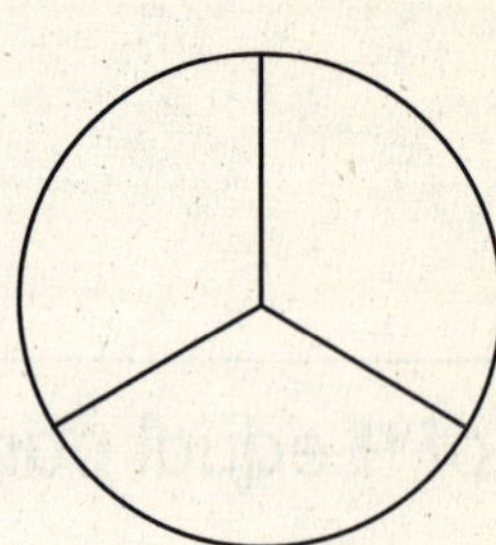

Circle the shape that shows fourths.

3.

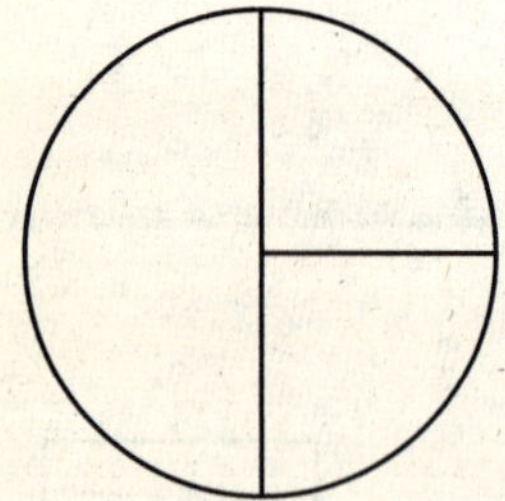

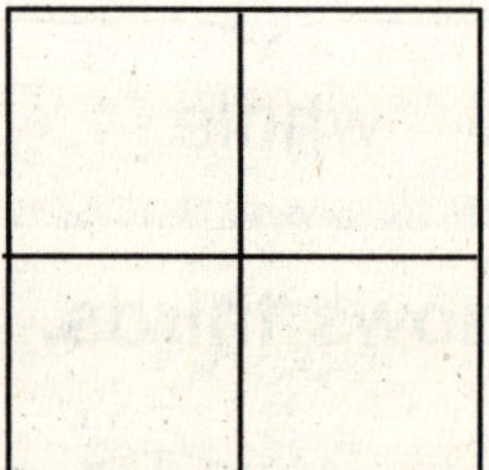

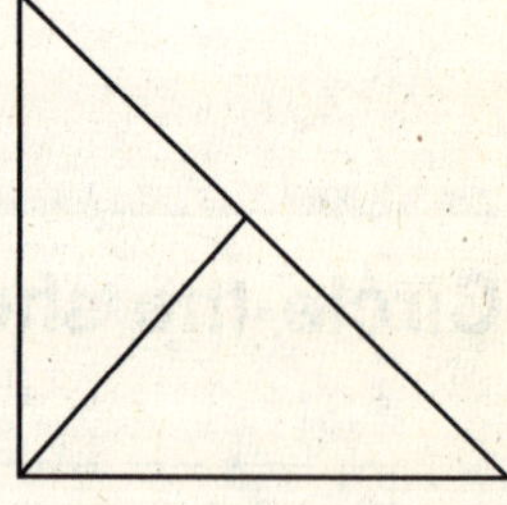

4.

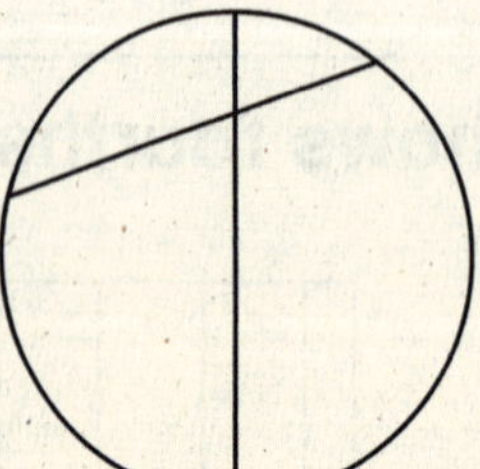

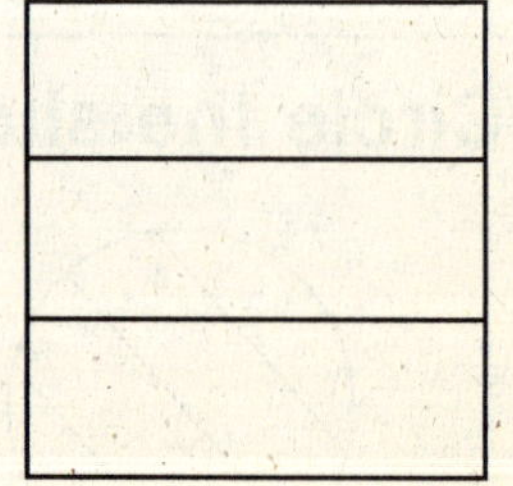

Name________________________________

Explore Place Value to 50

Skill 43

Learn the Math

You can use models to show numbers to 50.

Tens	Ones

34

3 tens 4 ones = 34

Vocabulary
tens
ones

Write how many tens and ones.
Write the number.

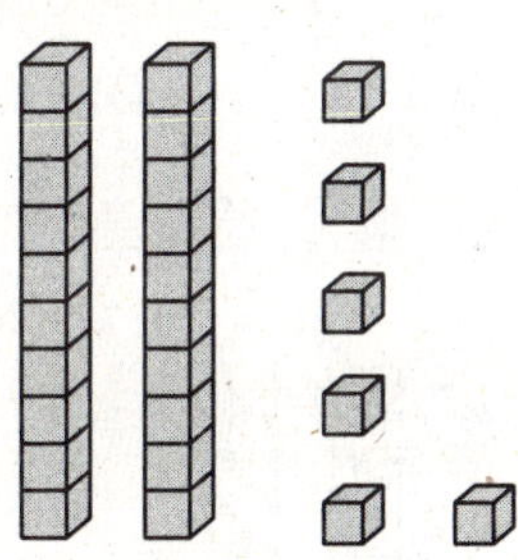

2 tens 6 ones = 26

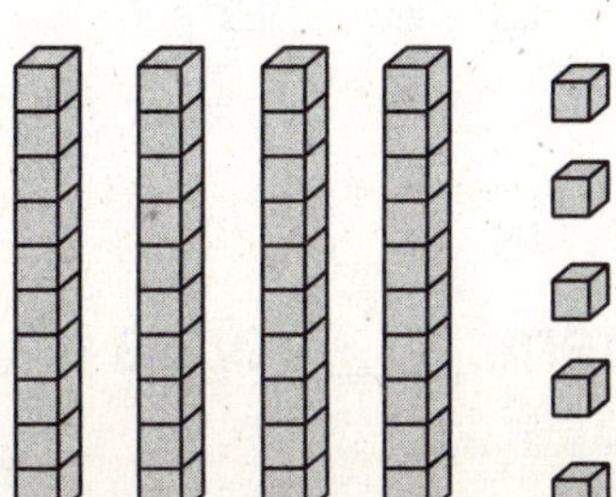

____ tens ____ ones = ____

Do the Math

Use [tens rod] and [ones cube].
Write how many tens and ones. Write the number.

1.
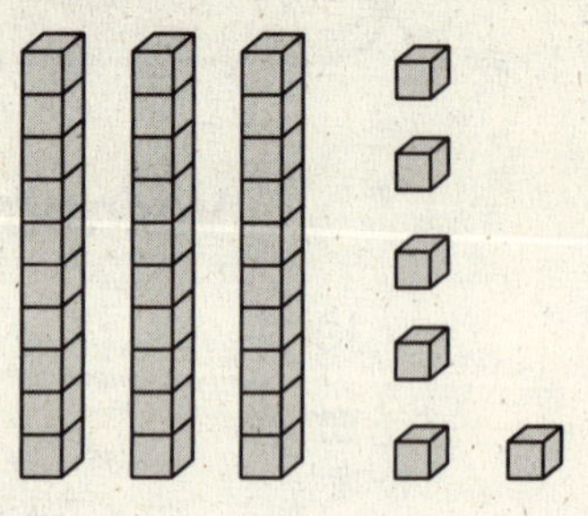

3 tens 6 ones = 36

2.
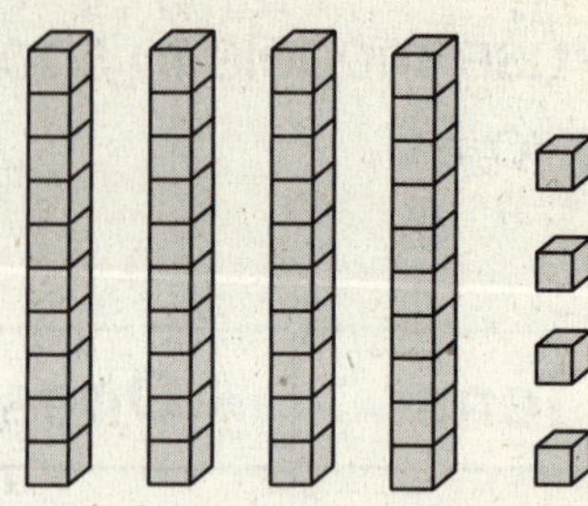

_____ tens _____ ones = _____

3.
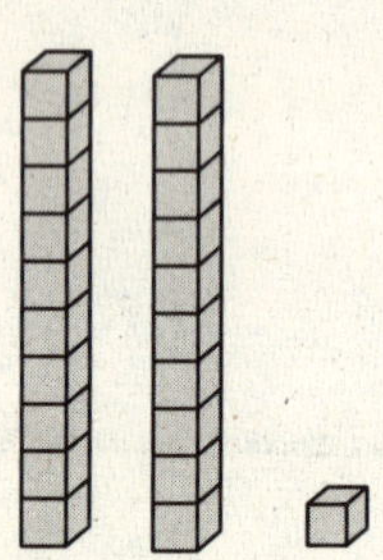

_____ tens _____ one = _____

4.
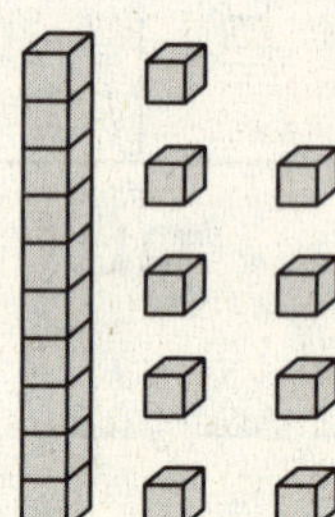

_____ ten _____ ones = _____

5.
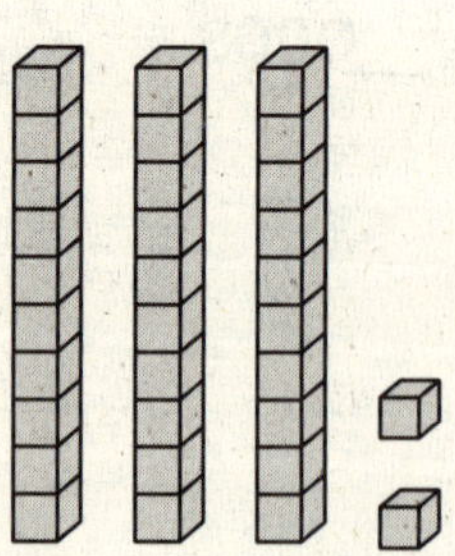

_____ tens _____ ones = _____

6.
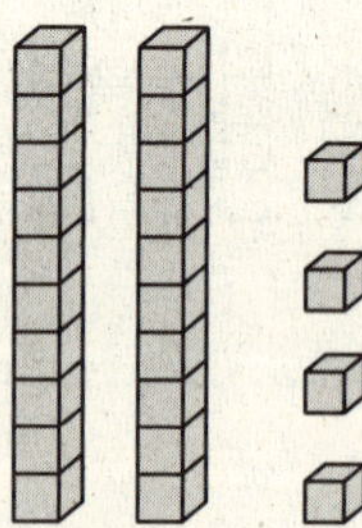

_____ tens _____ ones = _____

Name____________________

Explore Place Value to 100

Skill 44

Learn the Math

You can use models to show two-digit numbers to 100.

Vocabulary

tens
ones

Tens	Ones
(6 tens)	(2 ones)

62

6 tens 2 ones

62

Write how many tens and ones.
Write the number.

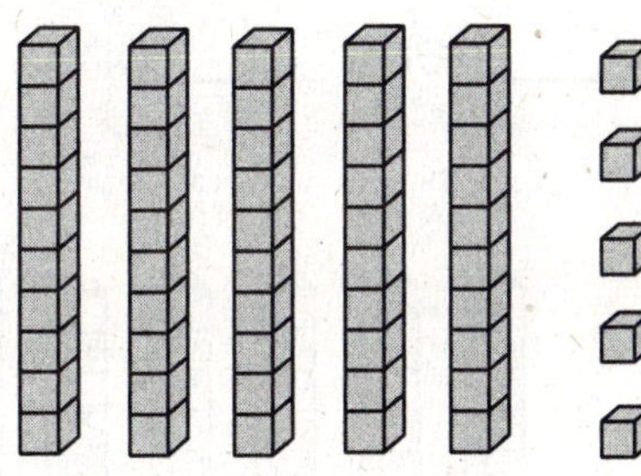

5 tens 5 ones

55

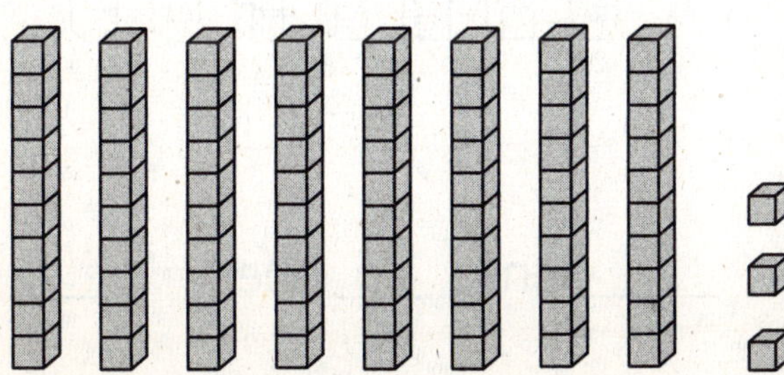

8 tens 3 ones

83

Use [ten-rod] and [unit cube].

Write how many tens and ones. Write the number.

1\.

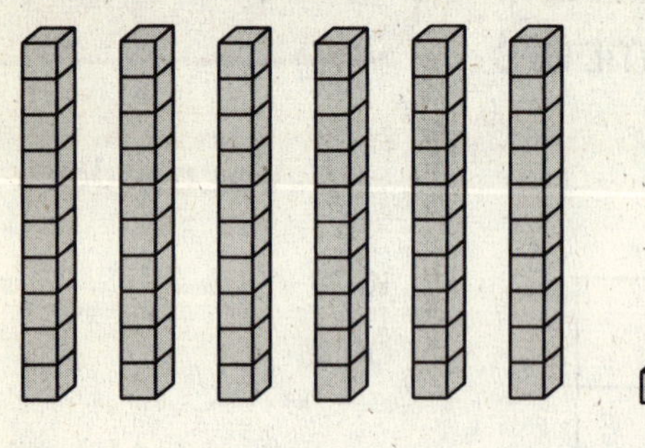

6 tens 1 one = 61

2\.

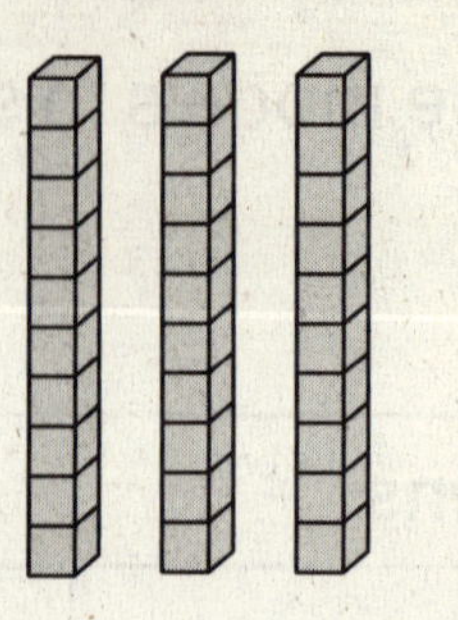

_____ tens _____ ones = _____

3\.

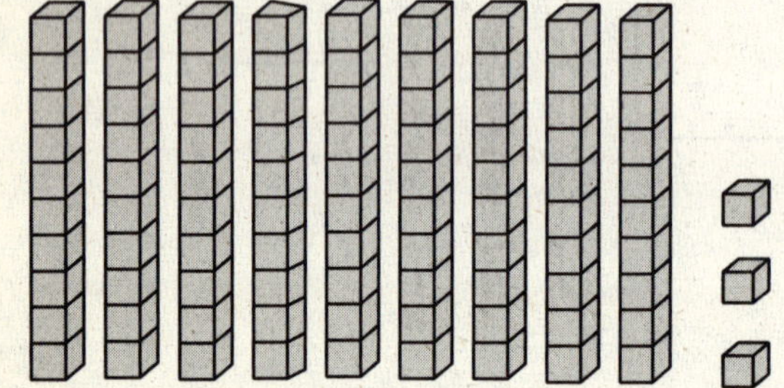

_____ tens _____ ones = _____

4\.

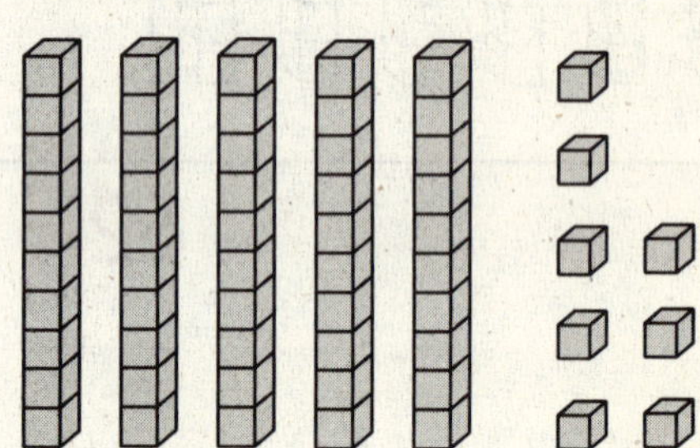

_____ tens _____ ones = _____

5\.

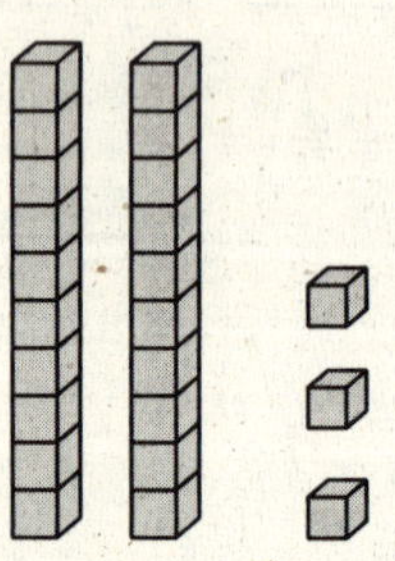

_____ tens _____ ones = _____

6\.

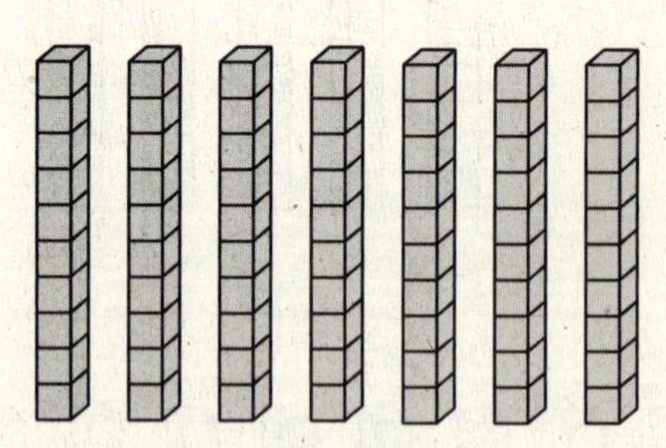

_____ tens _____ ones = _____

Name________________________________

Compare Numbers to 50

Skill 45

Learn the Math

You can use models to compare numbers.

Write the numbers. Circle the greater number.

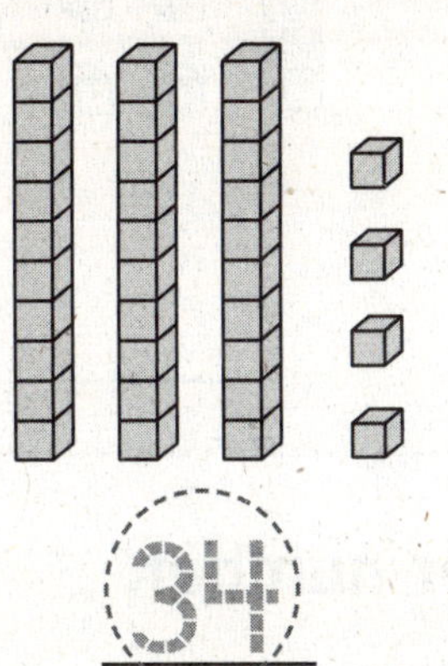

34

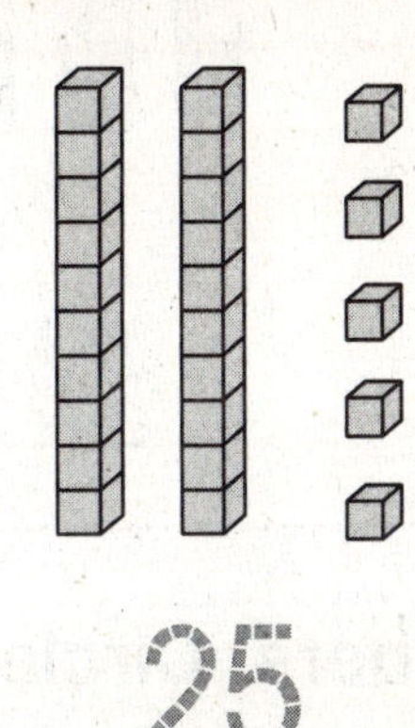

25

Write the numbers. Circle the lesser number.

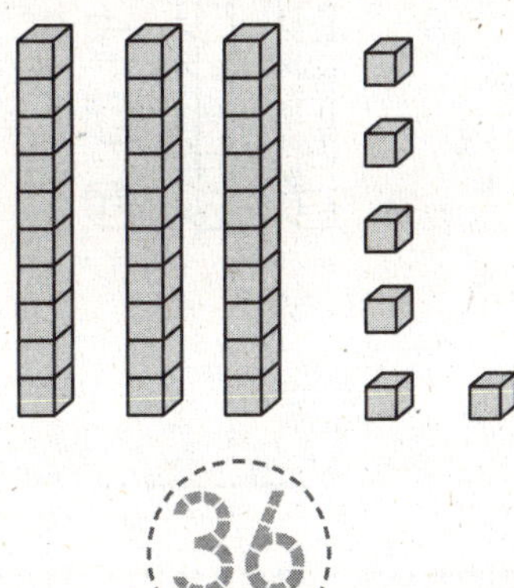

36

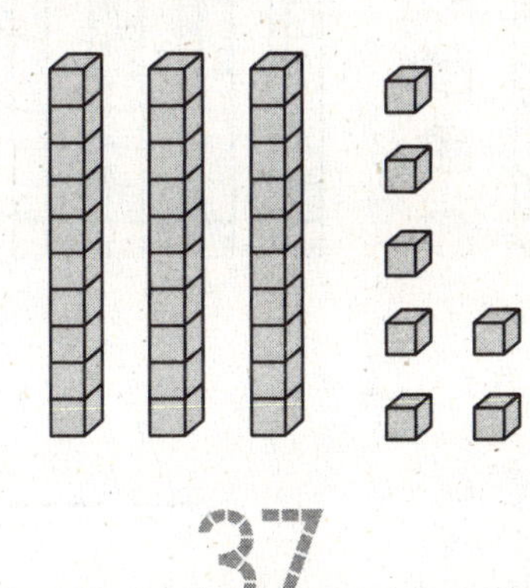

37

Write the numbers. Circle the numbers that are equal.

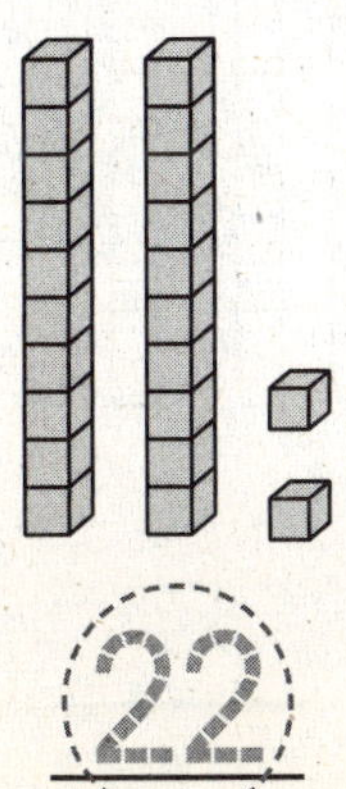

22

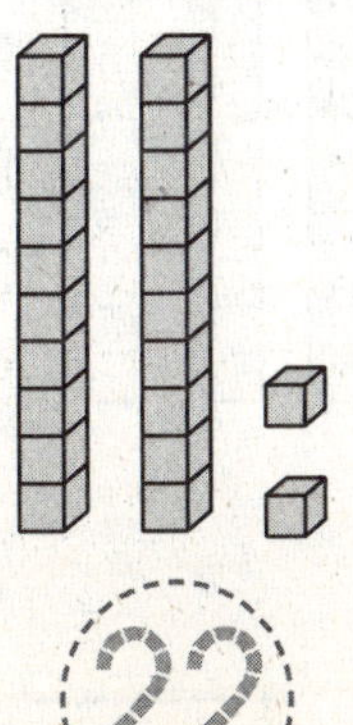

22

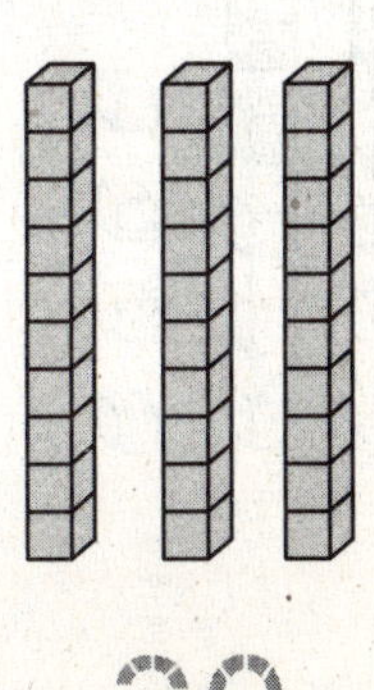

30

Write the numbers. Circle the greater number

1.

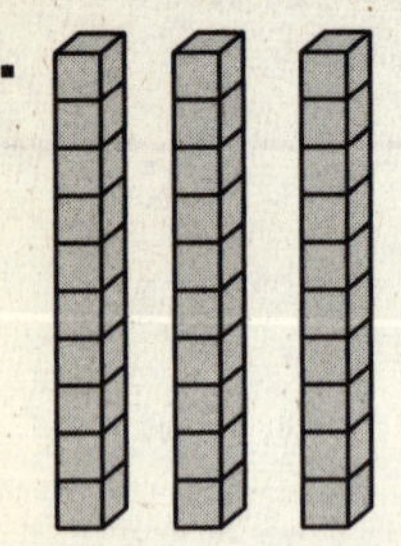

30

31

2.

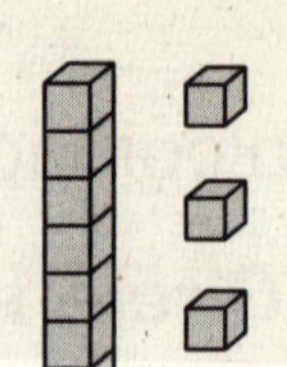

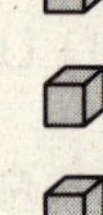

Write the numbers. Circle the lesser number.

3.

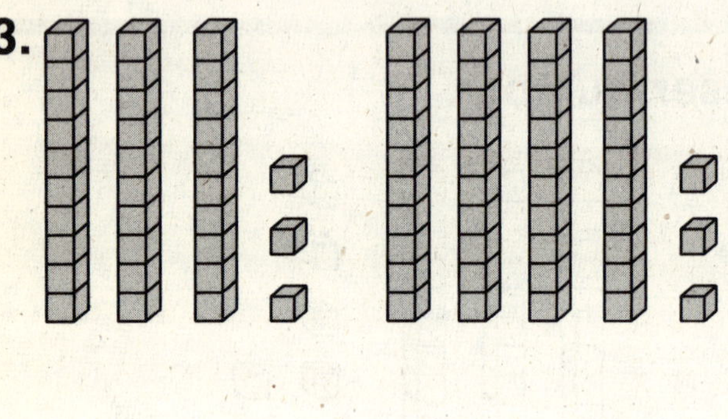

4.

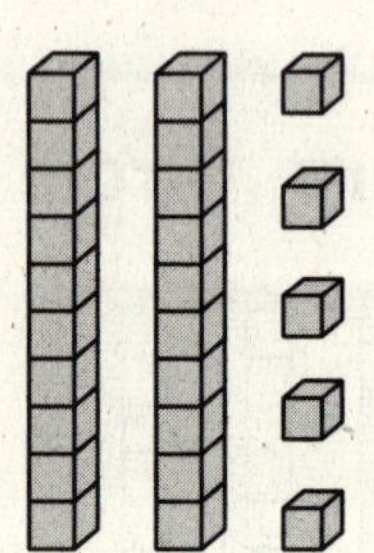

Write the numbers. Circle the numbers that are equal.

5.

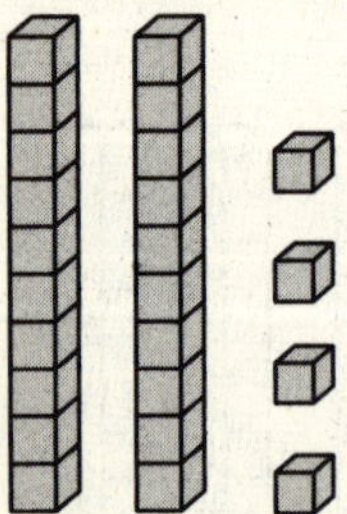

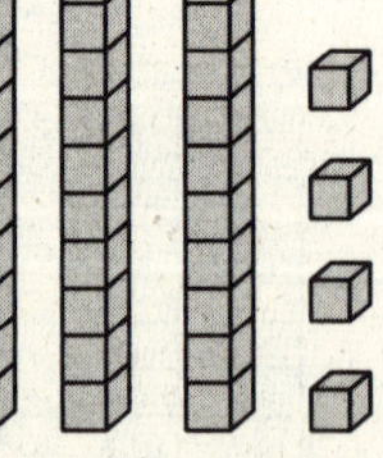

Name________________________________

Order Numbers to 100 on a Number Line

Skill 46

Learn the Math

You can use a number line to find the number that is before, between, or after other numbers.

Vocabulary

after
before
between

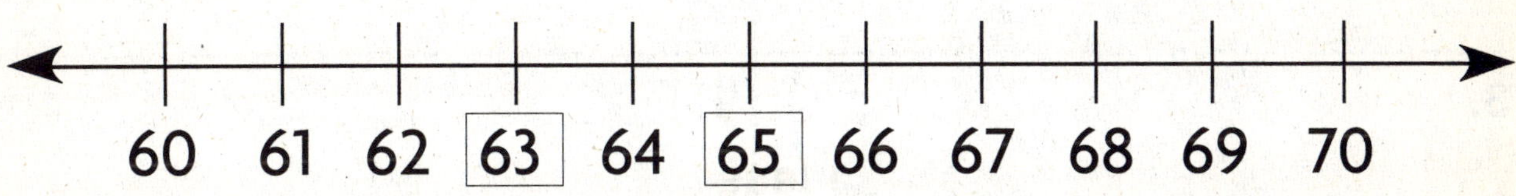

63 is just before 64.

64 is between 63 and 65.

65 is just after 64.

Write the number that is just before, between, or just after.

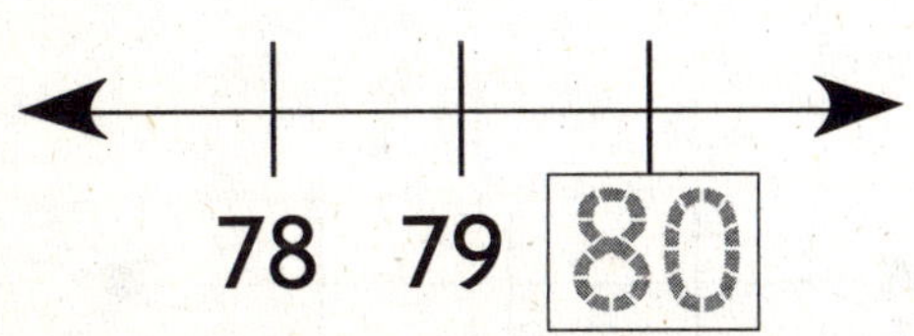

80 is just after 79.

Write the number that is just before, between, or just after.

1.

2.

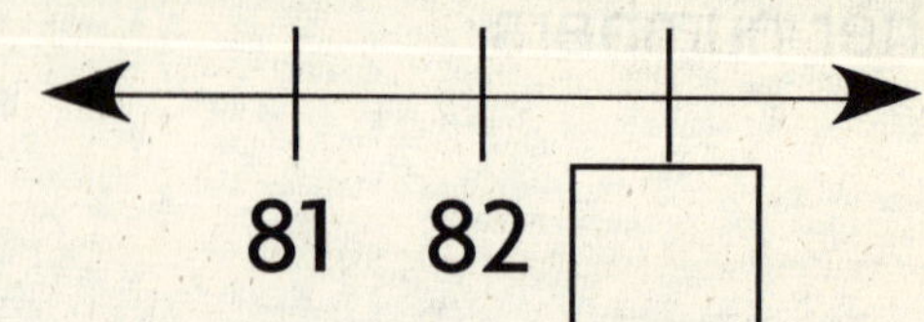

3.

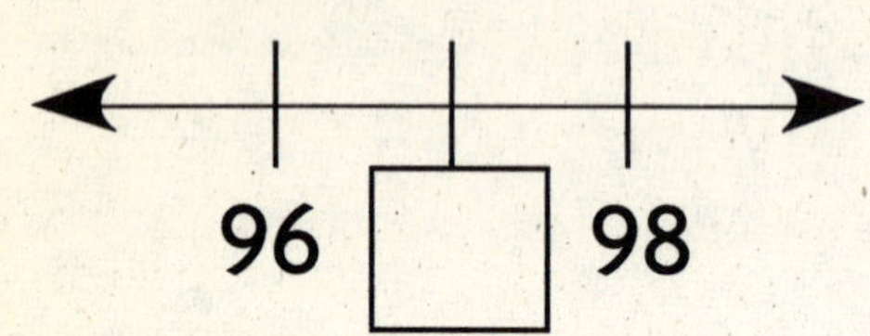

4.

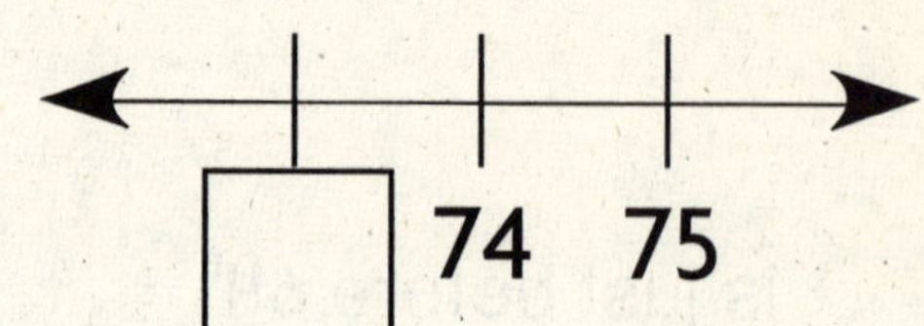

5.

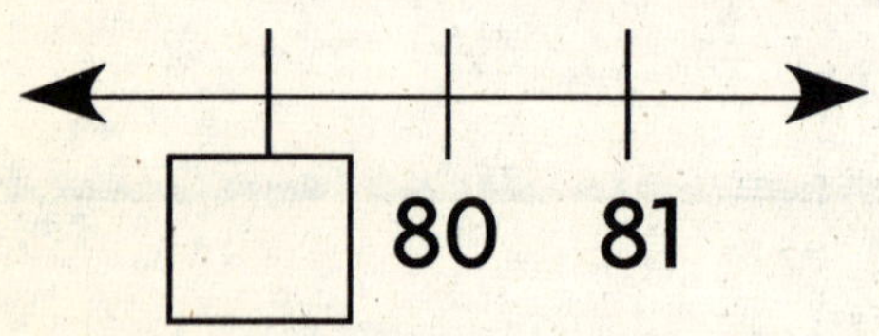

6.

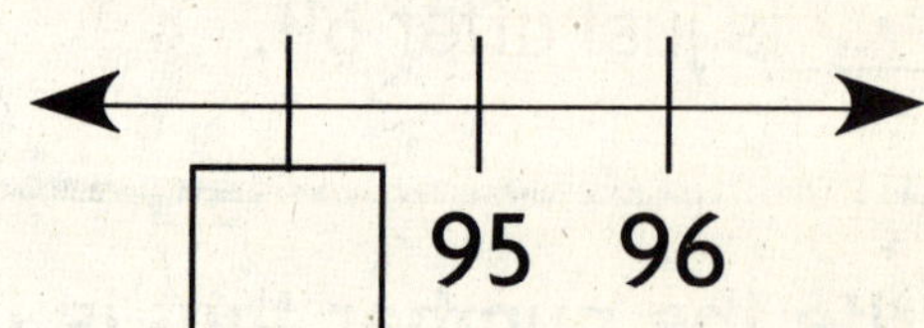

7.

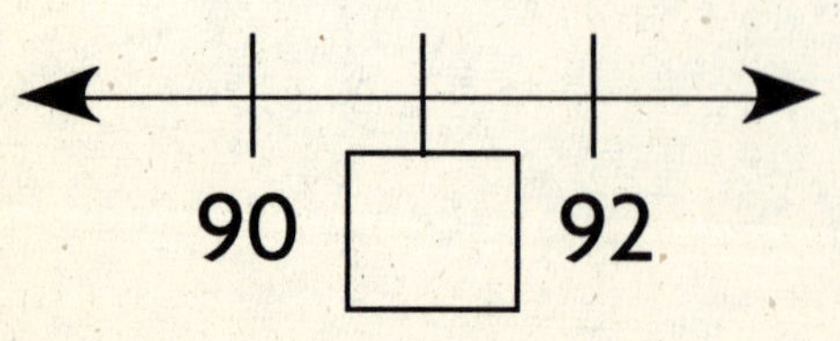

8.

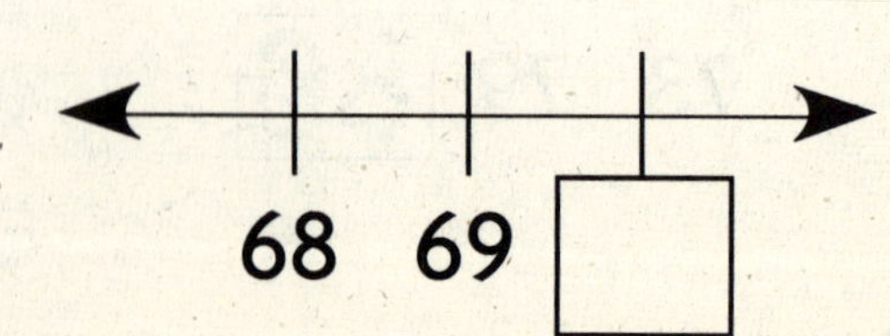

Name____

Model Addition with 1-Digit and 2-Digit Numbers

Skill 47

Learn the Math

You can use tens and ones to add.

Add 26 and 5.

Step 1 Show 26 and 5.

Tens	Ones

Step 2 Can you make a ten? If you can, regroup 10 ones for 1 ten.

Tens	Ones

Step 3 Write how many tens and ones. Write the sum.

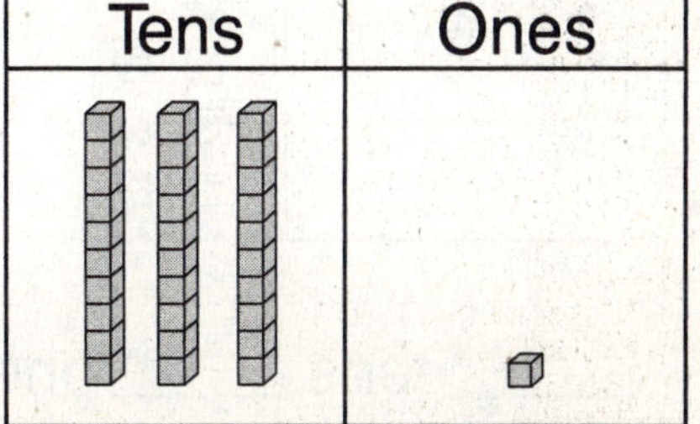

3 tens 1 one

______

Do the Math

Use [tens rod] [ones cube] to add. Write how many tens and ones. Write the sum.

1. Add 32 and 4.

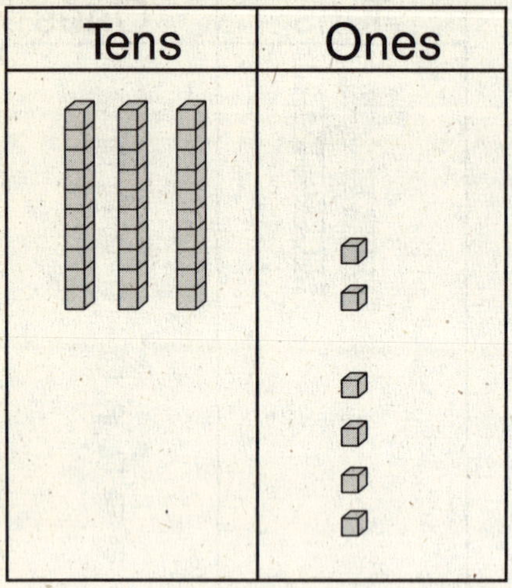

Tens	Ones

3 tens 6 ones

______

2. Add 27 and 6.

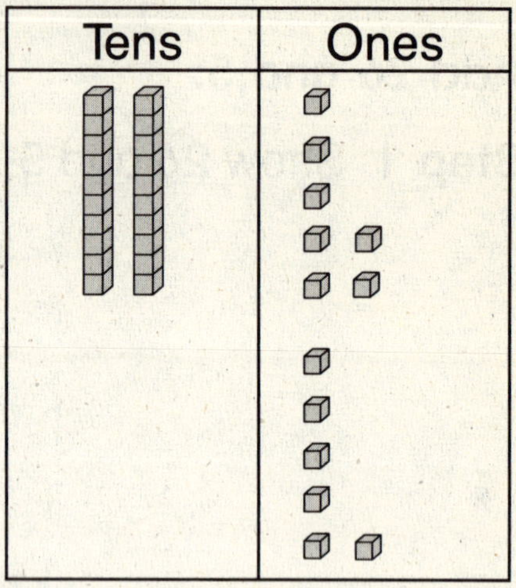

Tens	Ones

______ tens ______ ones

______

3. Add 41 and 8.

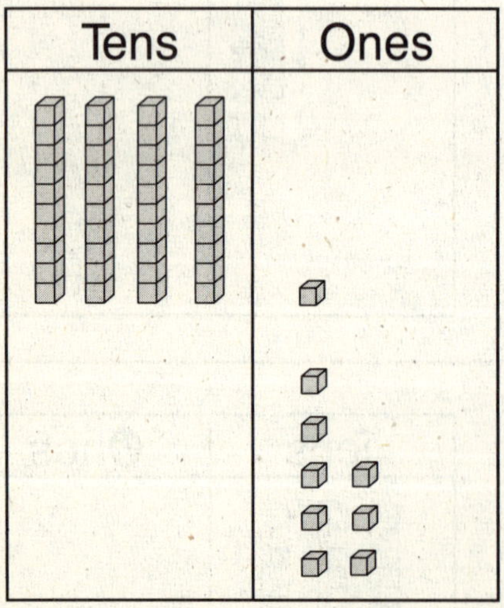

Tens	Ones

______ tens ______ ones

______

4. Add 34 and 7.

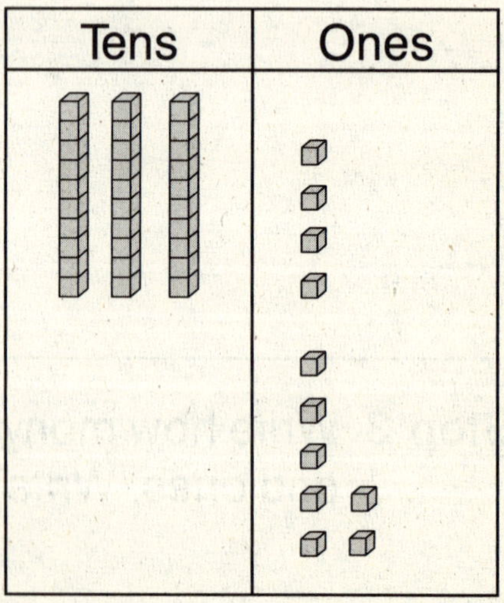

Tens	Ones

______ tens ______ one

______

Name________________________________

Mental Math: Add Tens

Skill 48

Learn the Math

You can use mental math to add tens.

50 + 20 = ?

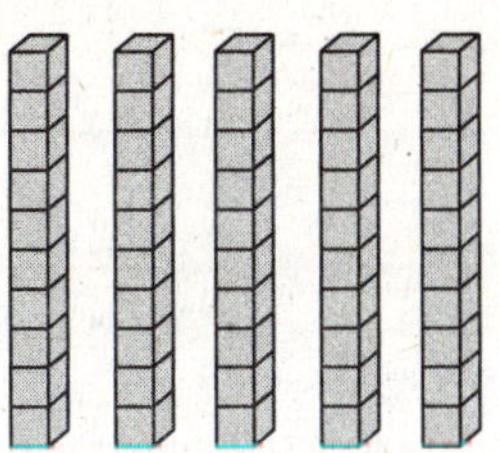

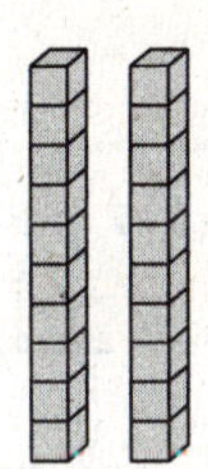

Think: 5 + 2 = 7

So, 5 tens + 2 tens = 7 tens.

7 tens is equal to 70.

So, 50 + 20 = 70 .

20 + 30 = ?

Think: 2 + 3 = 5

So, 2 tens + 3 tens = 5 tens.

5 tens is equal to ____.

So, 20 + 30 = ____ .

Add.

1. 4 + 3 = 7

4 tens + 3 tens = 7 tens

40 + 30 = 70

2. 5 + 4 = ___

5 tens + 4 tens = ___ tens

50 + 40 = ___

3. 2 + 6 = ___

2 tens + 6 tens = ___ tens

20 + 60 = ___

4. $\begin{array}{r} 30 \\ +\ 60 \\ \hline \end{array}$

5. $\begin{array}{r} 10 \\ +\ 10 \\ \hline \end{array}$

6. $\begin{array}{r} 20 \\ +\ 70 \\ \hline \end{array}$

7. $\begin{array}{r} 40 \\ +\ 40 \\ \hline \end{array}$

8. $\begin{array}{r} 60 \\ +\ 10 \\ \hline \end{array}$

9. $\begin{array}{r} 10 \\ +\ 20 \\ \hline \end{array}$

Name____

Model Subtraction with 2-Digit and 1-Digit Numbers

Skill 49

Learn the Math

You can use to subtract.

Subtract 8 from 36.

Step 1 Show 36.

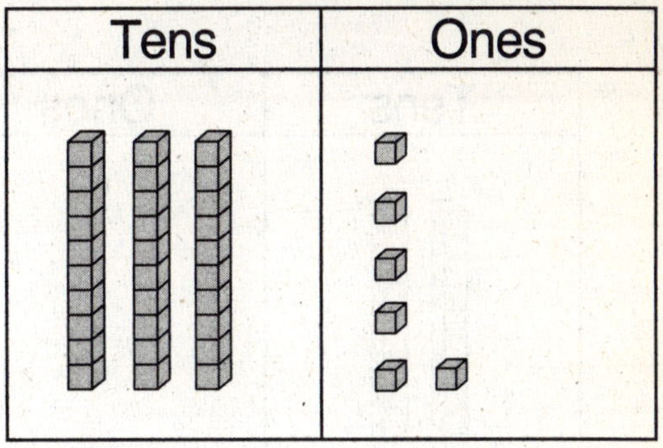

Step 2 Can you subtract 8 ones? If not, regroup 10 ones as 1 ten.

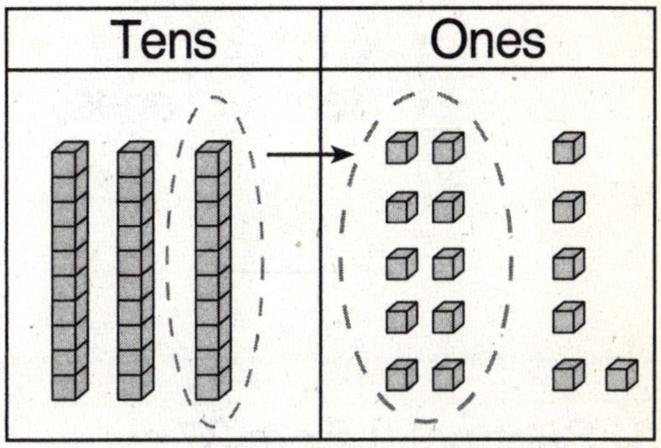

Step 3 Subtract 8 ones.

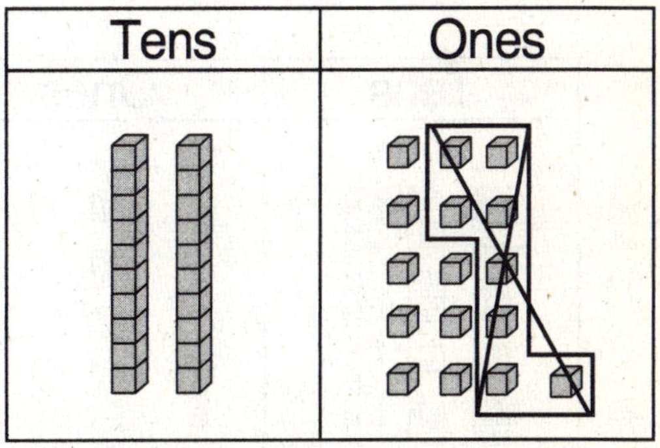

Step 4 Write how many tens and ones.

___ tens ___ ones

So, 8 from 36 is _____.

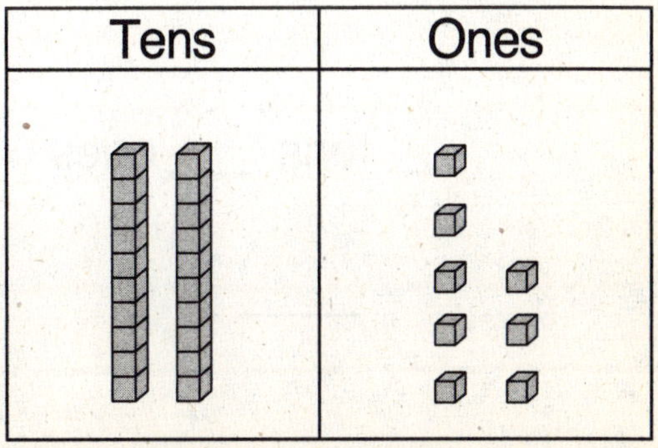

Use [tens block] [ones block] to subtract. Write how many tens and ones. Write the difference.

1. Subtract 5 from 21.

Tens	Ones

1 ten 6 ones

16

2. Subtract 3 from 46.

Tens	Ones

___ tens ___ ones

___

3. Subtract 4 from 37.

Tens	Ones

___ tens ___ ones

___

4. Subtract 7 from 34.

Tens	Ones

___ tens ___ ones

___

Name ______________________________

Mental Math: Subtract Tens

Skill 50

Learn the Math

You can use mental math to subtract tens.

60 − 40 = __?__

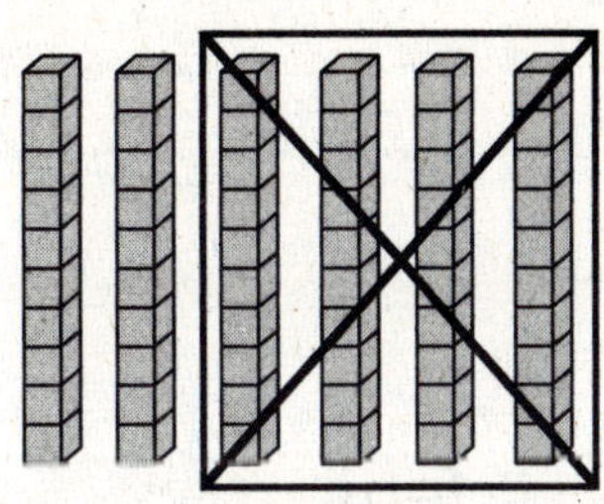

Think: 6 − 4 = 2

So, 6 tens − 4 tens = __2__ tens.

2 tens is equal to __20__.

So, 60 − 40 = __20__.

---

40 − 10 = __?__

Think: 4 − 1 = 3

So, 4 tens − 1 ten = __3__ tens.

3 tens is equal to ____ .

So, 40 − 10 = ____.

Subtract.

1. $7 - 5 =$ 2

7 tens − 5 tens = 2 tens

$70 - 50 =$ 20

2. $8 - 4 =$ ___

8 tens − 4 tens = ___ tens

$80 - 40 =$ ___

3. $5 - 3 =$ ___

5 tens − 3 tens = ___ tens

$50 - 30 =$ ___

4. $90 - 50 =$ ___

5. $70 - 30 =$ ___

6. $40 - 30 =$ ___

7. $50 - 10 =$ ___

8. $80 - 50 =$ ___

9. $90 - 70 =$ ___

Name____________________

Skip-Count by Twos and Fives

Skill 51

Learn the Math

You can skip-count by twos and fives.

Skip-count. Count the mittens by twos.

2 4 6 8 10

So, there are 10 mittens in all.

Skip-count. Count the fingers by fives.

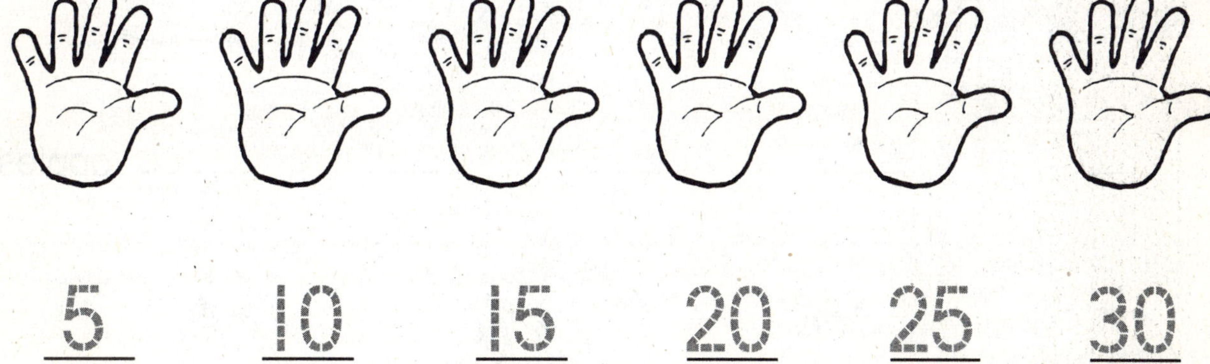

5 10 15 20 25 30

So, there are _____ fingers in all.

Skip-count by twos or fives. Write how many.

1.      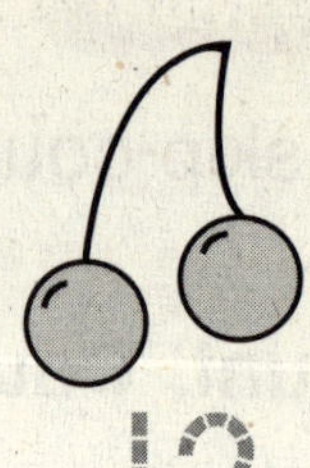

2 4 6 8 10 12 cherries

2.      

____ ____ ____ ____ ____ ____ flowers

3.     

____ ____ ____ ____ ____ apples

4.    

____ ____ ____ ____ bananas

Name ____________________

Skip-Count on a Hundred Chart

Skill 52

Learn the Math

You can use a hundred chart to skip-count.

1	2	3	4	5	6	7	8	9	10
11	12	13	14	15	16	17	18	19	20
21	22	23	24	25	26	27	28	29	30
31	32	33	34	35	36	37	38	39	40
41	42	43	44	45	46	47	48	49	50
51	52	53	54	55	56	57	58	59	60
61	62	63	64	65	66	67	68	69	70
71	72	73	74	75	76	77	78	79	80
81	82	83	84	85	86	87	88	89	90
91	92	93	94	95	96	97	98	99	100

Use the hundred chart to skip-count.
Start at 24. Skip-count by twos.

24, 26, 28, 30, 32, 34, 36

Start at 20. Skip-count by fives.

20, 25, 30, 35, 40, 45, 50

Do the Math

Use the hundred chart to skip-count.

1	2	3	4	5	6	7	8	9	10
11	12	13	14	15	16	17	18	19	20
21	22	23	24	25	26	27	28	29	30
31	32	33	34	35	36	37	38	39	40
41	42	43	44	45	46	47	48	49	50
51	52	53	54	55	56	57	58	59	60
61	62	63	64	65	66	67	68	69	70
71	72	73	74	75	76	77	78	79	80
81	82	83	84	85	86	87	88	89	90
91	92	93	94	95	96	97	98	99	100

1. Start at 15. Skip-count by fives.

15, 20, 25, 30, 35, 40, 45

2. Start at 42. Skip-count by twos.

42, ____, ____, ____, ____, ____, ____

3. Start at 30. Skip-count by fives.

30, ____, ____, ____, ____, ____, ____

4. Start at 20. Skip-count by tens.

20, ____, ____, ____, ____, ____, ____